Computational Psychiatry

Rodrick Wallace

Computational Psychiatry

A Systems Biology Approach to the Epigenetics of Mental Disorders

Springer

Rodrick Wallace
New York State Psychiatric Institute
New York, NY, USA

ISBN 978-3-319-53909-6 ISBN 978-3-319-53910-2 (eBook)
DOI 10.1007/978-3-319-53910-2

Library of Congress Control Number: 2017934644

Printed on acid-free paper

This Springer imprint is published by Springer Nature
The registered company is Springer International Publishing AG
The registered company address is: Gewerbestrasse 11, 6330 Cham, Switzerland

Preface

US psychiatry is in crisis. Perhaps the most evident symptom is the 'trans-institutionalization' of those with serious mental illness (SMI) from hospitals to jails and prisons, following decades of psychiatric bed eliminations. For New York State, some 96,000 such beds were available in the mid-1950s. About 3000 remain. New York City's Rikers Island jail houses some 11,000 inmates, with about 4000 now diagnosed as suffering SMI. Statewide, approximately 60% of hospital beds in jails and prisons are occupied by those with SMI. Current policy initiatives focus on making this mass jailing of the mentally ill 'more humane'. The pattern is repeated nationally. As Bastimpillai et al. (2016) note:

> Currently, the United States has a relatively low 22 psychiatric beds per 100,000 population compared with the Organisation for Economic Cooperation and Development (OECD) average of 71 beds per 100,000 population. Only 4 of the 35 OECD countries (Italy, Chile, Turkey, and Mexico) have fewer psychiatric beds per 100,000 population than the United States... Germany, Switzerland, and France have 127, 91, and 87 psychiatric beds per 100,000 population, respectively.
> (Bastiampillai, T., S. Sharfstein, S. Allison, 2016, US suicide rates and the critical decline in psychiatric beds,
> JAMA, doi: 10.1001/jama.2016.16989)

Further, the pharmaceutical industry has largely abandoned attempts to develop new drugs for SMI, beyond some continuing work on supposedly 'organic' dementias such as Alzheimer's disease, which are themselves highly stratified at the population level according to education, income and social status.

Intellectually, mainstream psychiatric doctrine exceeds even the discredited neoliberal atomism of mainstream economics, committing the mereological fallacy of attributing to 'the brain' what are complex disorders of not merely the entire individual but the individual enmeshed in both culture and an environment that, for humans, involves mostly other humans and their institutions.

Indeed, this latter circumstance has not gone entirely unremarked. As Huys et al. (2016) put it:

> ...[M]ental health depends not only on the function of the brain...but also on how that function relates to, influences, and is influenced by the individual's environmental and

> experiential challenges. Understanding mental health, and its disruption, therefore relies on linking multiple interacting levels, from molecules to cells, circuits, cognition, behavior, and the physical and social environment.
> (Huys, Q., T. Maia, M. Frank, 2016, Computational psychiatry as a bridge from neuroscience to clinical applications, Nature Neuroscience, 19: 404–413)

But, although there are many attempts in the literature to mathematically model neural process and, in some measure, its dysfunction, for ideological reasons, there is little actually available that places mental function within sociocultural, socioeconomic and environmental context. What does exist is widely scattered, both as peer-reviewed papers and individual chapters. This book brings the author's work in this direction into a single place and significantly extends it.

One central point is that a large class of cognitive processes can be approximated in terms of information sources. This is because cognition involves active choice of a response to impinging signals from a larger set of available alternatives. Choice reduces uncertainty, in a formal manner, and reduction in uncertainty implies the existence of an information source that we take as 'dual' to the cognitive process. The asymptotic limit theorems of information and control theories then impose themselves, allowing construction of necessary conditions statistical models, roughly as the central limit theorem allows construction of least squares regression models. While statistics is not science, statistical models permit comparison of similar systems under different conditions and different systems under similar conditions. To understate the case, iterating such comparisons can be of great use in refining scientific inference based on observational or empirical data.

A second central point is that, taking the perspective of Bennett and Feynman, in spite of its mathematical form, information is not an entropy, but a kind of free energy. Indeed, it is easy to construct an elegant little ideal machine that converts the information within a message to useful work. It then becomes possible to construct dynamic statistical models akin to Onsager's treatment of nonequilibrium thermodynamics, albeit without 'reciprocity relations' since most information sources are not even locally reversible: palindromes are very rare.

The first chapter examines the primary mental experience—consciousness—from an evolutionary perspective, recognizing the ubiquitous role played by the exaptation of crosstalk between cognitive modules at many different scales and levels of organization, in an explosion of parallel traits very similar to such examples as the many different forms of wings for flight.

Chapter 2 explores the missing heritability of complex diseases, focusing on 'cultural epigenetics' as essential contributing 'dark matter', while Chapter 3 makes the case that all forms of SMI are—necessarily and inherently—'culture-bound syndromes'. Chapter 4 introduces control theory tools to explore the environmental induction of neurodevelopmental disorders, in a large sense, and Chapter 5 continues with a study of the synergism between culture, psychopathology and sleep disorders.

Chapter 6 reexamines function and dysfunction from the perspective of embodiment. Chapter 7 introduces 'hidden symmetry' methods that should be useful in future studies of this nature. Chapters 8 and 9 use perspectives developed from computational psychiatry to examine failure modes of autonomous vehicle and autonomous/centaur weapon systems, emerging as de facto 'psychopathologies' of automata. Chapter 10 applies approaches from evolutionary economics to the self-referential dynamics of environmental insult, finding ratchet dynamics that can trigger increase or decline in factors producing developmental and cognitive dysfunction. The final chapter examines 'the madness of crowds' from the perspective of recent US security doctrine.

The chapters can be read separately according to interest; hence there is some considerable repetition of methodology between them, but many mathematical details have been collected into a general appendix.

Contrary, perhaps, to current expectations, it appears that the level of mathematics needed to address cognitive function and dysfunction in full context rivals, and likely exceeds, the norms of general relativity, high-energy particle physics, theoretical chemistry, and string theory.

About the Author

Rodrick Wallace has an undergraduate degree in mathematics and a Ph.D. in physics from Columbia University. He completed postdoctoral training in the epidemiology of mental disorders at Rutgers University and is a research scientist in the Division of Epidemiology of the New York State Psychiatric Institute. A past recipient of an Investigator Award in Health Policy Research from the Robert Wood Johnson Foundation, he is a former public interest lobbyist and the author of numerous peer-reviewed papers and books across a variety of disciplines. His work specializes in understanding the roles of state policy, historical trajectory, culture and socioeconomic structure in determining patterns of public health and order.

Two poems by Alfonz Wallace

WARNING

Look into his mind, and then
Never dare to look again;
For, in all that we might see,
There is much of you and me.

BLACK NARCISSUS

Of all night's strange inhabitants,
The creature I fear worst
Never betrays the countenance
That makes my sleep accursed.

I flee and search, finding no place
His dark shape will not find,
Who lives in my own body's space
And borrows my own mind.

Contents

Chapter 1
Consciousness, Crosstalk, and the Mereological Fallacy

Nothing in biology makes sense except in light of evolution.
– T. Dobzhansky

Summary The cross-sectional decontextualization afflicting contemporary neuroscience—attributing to "the brain" what is the province of the whole individual—is mirrored by an evolutionary decontextualization exceptionalizing consciousness. The living state is characterized by cognitive processes at all scales and levels of organization. Many can be associated with dual information sources that "speak" a "language" of behavior-in-context. Shifting, tunable, global broadcasts analogous to consciousness, albeit far slower—wound healing, tumor control, immune function, gene expression, etc.—have emerged through repeated evolutionary exaptation of the crosstalk and noise inherent to all information transmission. These broadcasts recruit "unconscious" cognitive modules into shifting arrays as needed to meet the threats and opportunities that confront all organisms across multiple frames of reference. The development is straightforward, based on the powerful necessary conditions imposed by the asymptotic limit theorems of communication theory, in the same sense that the Central Limit Theorem constrains sums of stochastic variates. Recognition of information as a form of free energy instantiated by physical processes that consume free energy permits analogs to phase transition and nonequilibrium thermodynamic arguments, leading to "dynamic regression models" useful for data analysis.

1.1 Introduction

The neuroscientist Max Bennett and the philosopher Peter Hacker characterize contemporary neuroscience, and particularly consciousness studies, as fatally contaminated by a decontextualization that attributes to "the brain" functions and dynamics inherent to the whole individual, what they call the "mereological fallacy" (Bennett and Hacker 2003). It is possible, with some formal overhead, to evade this version of the fallacy by erecting a more comprehensive scientific structure that embeds the shifting, tunable global neural broadcasts of individual animal consciousness (e.g., Baars 1988, 2005) within a nested hierarchy of cross-sectional physiological and sociocultural phenomena (e.g., Wallace 2005, 2007; Wallace and Fullilove 2008; Wallace and Wallace 2013).

R. Wallace, *Computational Psychiatry*, DOI 10.1007/978-3-319-53910-2_1

Unfortunately, another fatal manifestation of the fallacy is unaddressed. There remains the longitudinal decontextualization of consciousness from the body of evolutionary theory that is the foundation of contemporary biology. Was consciousness, then, sprung Athena-like and full-armed from the brow of an adaptationist Zeus some 500 million years ago? A rather different kind of "just so" story is needed that recognizes the necessary conditions imposed by the asymptotic limit theorems of communication theory on all biological and ecological processes involving the production and transmission of information.

Recall that information sources produce, and channels transmit, information, and that transmission and production are constrained by the Shannon Coding, the Shannon–McMillan Source Coding, and the Rate Distortion Theorems (e.g., Khinchin 1957; Ash 1990; Cover and Thomas 2006; Shannon 1959). These are as confining as the Central Limit Theorem is for sums of independent stochastic variates, as the Martingale Theorem is for repeated games of chance, and as the Ergodic Theorem is in converging time averages to cross-sectional means for a broad class of probabilistic phenomena.

Indeed, parallel to the neuroscience and philosophy debate is what Adams (2003) calls "the informational turn in philosophy"—explicit application of communication theory formalism and concepts to "purposive behavior, learning, pattern recognition, and... the naturalization of mind and meaning." One of the first comprehensive attempts was that of Dretske (1981, 1988, 1992, 1993, 1994), whose work Adams describes:

> It is not uncommon to think that information is a commodity generated by things with minds. Let's say that a naturalized account puts matters the other way around, viz. it says that minds are things that come into being by purely natural causal means of exploiting the information in their environments. This is the approach of Dretske as he tried consciously to unite the cognitive sciences around the well-understood mathematical theory of communication...

Dretske himself (1994) writes:

> Communication theory can be interpreted as telling one something important about the conditions that are needed for the transmission of information as ordinarily understood, about what it takes for the transmission of semantic information. This has tempted people... to exploit [information theory] in semantic and cognitive studies, and thus in the philosophy of mind.
>
> ...Unless there is a statistically reliable channel of communication between [a source and a receiver]... no signal can carry semantic information... [thus] the channel over which the [semantic] signal arrives [must satisfy] the appropriate statistical constraints of communication theory.

It is fruitful to redirect attention from the informational content or meaning of individual symbols—the province of semantics which so concerned Dretske—back to the statistical properties of long, internally structured strings of signals emitted by an information source that is "dual" in a certain formal manner to a cognitive process. Application of a variety of tools adapted from statistical physics produces a dynamically tunable punctuation or phase transition coupling interacting cognitive modules in a highly natural way.

As Dretske so clearly saw, this approach allows scientific inference on the necessary conditions for cognition, greatly illuminating formal models of consciousness and other processes within and across organisms without raising the eighth Century ghosts of noisy, distorted mechanical clocks inherent to dynamic systems theory. The technique permits extension far beyond what is possible from statistical mechanics treatments of neural networks. In essence, the method broadly recapitulates the General Linear Model (GLM) for independent or simple serially correlated observations, but on the far more complex behaviors of an interacting assembly of information sources, using the Shannon Coding, Shannon–McMillan Source Coding, and Rate Distortion Theorems rather than the Central Limit Theorem.

Later chapters will explore the implications of the Data Rate Theorem that links control and information theories.

Wallace (2010) has examined the central role of information transmission constraints in evolutionary process, but, going in the opposite direction, this analysis will show how the inherent "flaws" of noise and crosstalk provide hooks for exaptations, in the sense of Gould (2002), that enabled evolution of a broad spectrum of tunable physiological global broadcasts, of which consciousness is merely a late, if remarkably rapid, example.

The next sections provide case histories of phenomena that recruit underlying sets of cognitive processes—"unconscious" modules—into a larger whole, including the Baars model of consciousness itself. The various examples are, however, much in the spirit of Maturana and Varela (1980, 1992) who long ago understood the central role that cognition must play in biological phenomena.

1.2 Some Cognitive Global Broadcasts

Immune System

Atlan and Cohen (1998) have proposed an information-theoretic—and implicitly global broadcast—cognitive model of immune function and process, a paradigm incorporating cognitive pattern recognition-and-response behaviors that are certainly analogous to, but much slower than, those of the later-evolved central nervous system.

From the Atlan/Cohen perspective, the meaning of an antigen can be reduced to the type of response the antigen generates. That is, the meaning of an antigen is functionally defined by the response of the immune system. The meaning of an antigen to the system is discernible in the type of immune response produced, not merely whether or not the antigen is perceived by the receptor repertoire. Because the meaning is defined by the type of response there is indeed a response repertoire and not only a receptor repertoire.

To account for immune interpretation, Cohen (1992, 2000) has reformulated the cognitive paradigm for the immune system. The immune system can respond to a given antigen in various ways, it has "options." Thus the particular response observed is the outcome of internal processes of weighing and integrating information about the antigen.

In contrast to Burnet's view of the immune response as a simple reflex, it is seen to exercise cognition by the interpolation of a level of information processing between the antigen stimulus and the immune response. A cognitive immune system organizes the information borne by the antigen stimulus within a given context and creates a format suitable for internal processing; the antigen and its context are transcribed internally into the "chemical language" of the immune system.

The cognitive paradigm suggests a language metaphor to describe immune communication by a string of chemical signals. This metaphor is apt because the human and immune languages can be seen to manifest several similarities such as syntax and abstraction. Syntax, for example, enhances both linguistic and immune meaning.

Although individual words and even letters can have their own meanings, an unconnected subject or an unconnected predicate will tend to mean less than does the sentence generated by their connection.

The immune system creates a "language" by linking two ontogenetically different classes of molecules in a syntactical fashion. One class of molecules are the T and B cell receptors for antigens. These molecules are not inherited, but are somatically generated in each individual. The other class of molecules responsible for internal information processing is encoded in the individual's germline.

Meaning, the chosen type of immune response, is the outcome of the concrete connection between the antigen subject and the germline predicate signals.

The transcription of the antigens into processed peptides embedded in a context of germline ancillary signals constitutes the functional language of the immune system. Despite the logic of clonal selection, the immune system does not respond to antigens as they are, but to abstractions of antigens-in-context, and does so in a dynamic manner across many tissue—global broadcasts.

Tumor Control

Nunney (1999) has explored cancer occurrence as a function of animal size, suggesting that in larger animals, whose lifespan grows as about the 4/10 power of their cell count, prevention of cancer in rapidly proliferating tissues becomes more difficult in proportion to size. Cancer control requires the development of additional mechanisms and systems to address tumorigenesis as body size increases—a synergistic effect of cell number and organism longevity. Nunney concludes that this pattern may represent a real barrier to the evolution of large, long-lived animals and predicts that those that do evolve have recruited additional controls over those of smaller animals to prevent cancer.

In particular, different tissues may have evolved markedly different tumor control strategies. All of these, however, are likely to be energetically expensive, permeated with different complex signaling strategies, and subject to a multiplicity of reactions to signals, including those related to psychosocial stress. Forlenza and Baum (2000) explore the effects of stress on the full spectrum of tumor control, ranging from DNA damage and control, to apoptosis, immune surveillance, and mutation rate. R. Wallace et al. (2003) argue that this elaborate tumor control strategy, in large animals, must be at least as cognitive as the immune system itself, one of its principal components: some comparison must be made with an internal picture of a healthy cell, and a choice made as to response, i.e., none, attempt DNA repair, trigger programmed cell death, engage in full-blown immune attack. This is, from the Atlan/Cohen perspective, the essence of cognition, and clearly involves the recruitment of a comprehensive set of cognitive subprocesses into a larger, highly tunable, dynamic structure.

Wound Healing

Following closely Mindwood et al. (2004), mammalian tissue repair is a series of overlapping events that begins immediately after wounding, as in Fig. 1.1 (Haggstrom 2012).

Platelet aggregation forms a hemostatic plug and blood coagulation forms the provisional matrix. This dense cross-linked network of fibrin and fibronectin from blood acts to prevent excessive blood loss. Platelets release growth factors and adhesive proteins that stimulate the inflammatory response, entraining immune function, and inducing cell migration into the wound using the provisional matrix as a substrate. Wound cleaning is done by neutrophils, solubilizing debris, and monocytes that differentiate into macrophages and phagocytose debris. The macrophages release growth factors and cytokines that activate subsequent events. For cutaneous wounds, keratinocytes migrate across the area to reestablish the epithelial barrier. Fibroblasts then enter the wound to replace the provisional matrix with granulation tissue composed of fibronectin and collagen. As endothelial cells revascularize the damaged area, fibroblasts differentiate into myofibroblasts and contract the

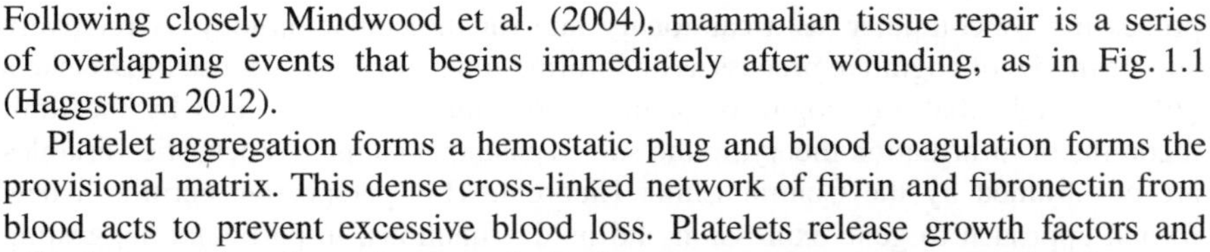

Fig. 1.1 Stages of wound healing on a logarithmic timescale, after Haggstrom (2012). Multiple subprocesses are recruited to the wound site in sequence, a global broadcast of cognitive phenomena slower, but no less sophisticated, than animal consciousness

matrix to bring the margins of the wound together. The resident cells then undergo apoptosis, leaving collagen-rich scar tissue that is slowly remodeled in the following months. Wound healing, then, provides an ancient example of a global broadcast that recruits a set of cognitive processes, in the sense of Atlan and Cohen (1998). The mechanism, which may vary across taxa, is inherently tunable, addressing the signal of "excessive distortion" represented by a wound.

Gene Expression

A cognitive paradigm for gene expression has emerged, a model in which contextual factors determine the behavior of what must be characterized as a "reactive system," not at all a deterministic—or even simple stochastic—mechanical process (Cohen 2006; Cohen and Harel 2007; Wallace and Wallace 2008, 2009, 2010).

O'Nuallain (2008) puts gene expression directly in the realm of complex linguistic behavior, for which context imposes meaning. He claims that the analogy between gene expression and language production is useful both as a fruitful research paradigm and also, given the relative lack of success of natural language processing by computer, as a cautionary tale for molecular biology. A relatively simple model of cognitive process as an information source permits use of Dretske's (1994) insight that any cognitive phenomenon must be constrained by the limit theorems of information theory, in the same sense that sums of stochastic variables are constrained by the Central Limit Theorem. This perspective permits a new formal approach to gene expression and its dysfunctions, in particular suggesting new and powerful statistical tools for data analysis that could contribute to exploring both ontology and its pathologies. Wallace and Wallace (2009, 2010) apply the perspective, respectively, to infectious and chronic disease.

This approach is consistent with the broad context of epigenetics and epigenetic epidemiology. Jablonka and Lamb (1995, 1998), for example, argue that information can be transmitted from one generation to the next in ways other than through the base sequence of DNA. It can be transmitted through cultural and behavioral means in higher animals, and by epigenetic means in cell lineages. All of these transmission systems allow the inheritance of environmentally induced variation. Such Epigenetic Inheritance Systems are the memory systems that enable somatic cells of different phenotypes but identical genotypes to transmit their phenotypes to their descendants, even when the stimuli that originally induced these phenotypes are no longer present.

After much research and debate, this epigenetic perspective has received much empirical confirmation (e.g., Backdahl et al. 2009; Turner 2000; Jaenisch and Bird 2003; Jablonka 2004).

Foley et al. (2009) argue that epimutation is estimated to be 100 times more frequent than genetic mutation and may occur randomly or in response to the environment. Periods of rapid cell division and epigenetic remodeling are likely to be most sensitive to stochastic or environmentally mediated epimutation. Disruption of epigenetic profile is a feature of most cancers and is speculated to play a role in the etiology of other complex diseases including asthma, allergy, obesity, type 2 diabetes, coronary heart disease, autism spectrum and bipolar disorders, and schizophrenia.

Scherrer and Jost (2007a,b) explicitly invoke information theory in their extension of the definition of the gene to include the local epigenetic machinery, a construct they term the "genon." Their central point is that coding information is not simply contained in the coded sequence, but is, in their terms, *provided by* the genon that accompanies it on the expression pathway and controls in which peptide it will end up. In their view the information that counts is not about the identity of a nucleotide or an amino acid derived from it, but about the relative frequency of the transcription and generation of a particular type of coding sequence that then contributes to the determination of the types and numbers of functional products derived from the DNA coding region under consideration.

The genon, as Scherrer and Jost describe it, is precisely a localized form of global broadcast linking cognitive regulatory modules to direct gene expression in producing the great variety of tissues, organs, and their linkages that comprise a living entity.

The proper formal tools for understanding phenomena that "provide" information—that are information sources—are the Rate Distortion Theorem and its zero error limit, the Shannon–McMillan Theorem.

Sociocultural Cognition

Humans are particularly noted for a hypersociality that inevitably enmeshes us all in group decisions and collective cognitive behavior within a social network, tinged by an embedding shared culture. For humans, culture is truly fundamental. Durham (1991), Richerson and Boyd (2006), Jablonka and Lamb (1995), and many others argue that genes and culture are two distinct but interacting systems of inheritance within human populations. Information of both kinds has influence, actual or potential, over behaviors, creating a real and unambiguous symmetry between genes and phenotypes on the one hand, and culture and phenotypes on the other. Genes and culture are best represented as two parallel lines or tracks of hereditary influence on phenotypes.

Much of hominid evolution can be characterized as an interweaving of genetic and cultural systems. Genes came to encode for increasing hypersociality, learning, and language skills. The most successful populations displayed increasingly complex structures that better aided in buffering the local environment (e.g., Bonner 1980).

Successful human populations seem to have a core of tool usage, sophisticated language, oral tradition, mythology, music, and decision-making skills focused on relatively small family/extended family social network groupings. More complex social structures are built on the periphery of this basic object (e.g., Richerson and Boyd 2006). The human species' very identity may rest on its unique evolved capacities for social mediation and cultural transmission. These are particularly expressed through the cognitive decision making of small groups facing changing patterns of threat and opportunity, processes in which we are all embedded and all participate.

The emergent cognitive behavior of organizations has, in fact, long been explored under the label "distributed cognition." As Hollan et al. (2000) describe,

> The theory of distributed cognition, like any cognitive theory, seeks to understand the organization of cognitive systems. Unlike traditional theories, however, it extends the reach of what is considered *cognitive* beyond the individual to encompass interactions between people and with resources and materials in the environment. It is important from the outset to understand that distributed cognition refers to a perspective on all of cognition, rather than a particular kind of cognition... Distributed cognition looks for cognitive processes, wherever they may occur, on the basis of the functional relationships of elements that participate together in the process. A process is not cognitive simply because it happens in a brain, nor is a process noncognitive simply because it happens in the interactions between many brains... In distributed cognition one expects to find a system that can dynamically configure itself to bring subsystems into coordination to accomplish various functions.

Wallace and Fullilove (2008) apply several of the formal models explored below to institutions and other social structures, and Wallace (2006, 2008, 2009, 2010, 2017) uses them to analyze canonical and idiosyncratic failure modes of massively parallel real-time computing systems.

1.3 Animal Consciousness

Sergent and Dehaene (2004) describe the context surrounding consciousness studies as follows:

> [A growing body of empirical work shows] large all-or-none changes in neural activity when a stimulus fails to be [consciously] reported as compared to when it is reported... [A] qualitative difference between unconscious and conscious processing is generally expected by theories that view recurrent interactions between distant brain areas as a necessary condition for conscious perception... One of these theories [that of Bernard Baars] has proposed that consciousness is associated with the interconnection of multiple areas processing a stimulus by a [dynamic] "neuronal workspace" within which recurrent connections allow long-distance communication and auto-amplification of the activation. Neuronal network simulations... suggest the existence of a fluctuating dynamic threshold. If the primary activation evoked by a stimulus exceeds this threshold, reverberation takes place and stimulus information gains access, through the workspace, to a broad range of [other brain] areas allowing, among other processes, verbal report, voluntary manipulation, voluntary action and long-term memorization. Below this threshold, however, stimulus information remains unavailable to these processes. Thus the global neuronal workspace theory predicts an all-or-nothing transition between conscious and unconscious perception... More generally, many non-linear dynamical systems with self-amplification are characterized by the presence of discontinuous transitions in internal state...

Thus Bernard Baars' global workspace model of animal consciousness sees the phenomenon as a dynamic array of unconscious cognitive modules that unite to become a global broadcast having a tunable perception threshold not unlike a theater spotlight, but whose range of attention is constrained by embedding contexts (Baars 1988, 2005). Baars and Franklin (2003) describe these matters as follows:

1. The brain can be viewed as a collection of distributed specialized networks (processors).
2. Consciousness is associated with a global workspace in the brain – a fleeting memory capacity whose focal contents are widely distributed—"broadcast"—to many unconscious specialized networks.
3. Conversely, a global workspace can also serve to integrate many competing and cooperating input networks.
4. Some unconscious networks, called contexts, shape conscious contents, for example, unconscious parietal maps modulate visual feature cells that underlie the perception of color in the ventral stream.
5. Such contexts work together jointly to constrain conscious events.
6. Motives and emotions can be viewed as goal contexts.
7. Executive functions work as hierarchies of goal contexts.

The basic mechanism emerges from a relatively simple application of the asymptotic limit theorems of information theory, once a broad range of unconscious cognitive processes is recognized as inherently characterized by information sources—generalized languages (Wallace 2000, 2005, 2007). This permits mapping physiological unconscious cognitive modules onto an abstract network of interacting information sources, allowing a simplified mathematical attack that, in the presence of sufficient linkage—crosstalk, permits rapid, shifting, global broadcasts in response to sufficiently large impinging signals. The topology of that broadcast is tunable, depending on the spectrum of distortion measure and other limits imposed on the system of interest.

While the mathematical description of consciousness is itself relatively simple, the evolutionary trajectories leading to its emergence seem otherwise. Here we argue that this is not the case, and that physical restrictions on the availability of metabolic free energy provide sufficient conditions for the emergence, not only of consciousness, but also of a spectrum of analogous "global" broadcast phenomena acting across a variety of biological scales of space, time, and levels of organization.

The argument is, in a sense, an inversion of Gould and Lewontin's (1979) famous essay "The Spandrels of San Marco and the Panglossian Paradigm: A Critique of the Adaptationist Programme." Spandrels are the triangular sectors of the intersecting arches that support a cathedral roof—simple byproducts of the need for arches—and their occurrence is in no way fundamental to the construction of a cathedral. Crosstalk between "low level" cognitive biological modules is a similar inessential by product that evolutionary process has exapted to construct the dynamic global broadcasts of consciousness and a spectrum of roughly analogous physiological phenomena: Evolution built many new arches from a single spandrel.

A formal overview, much like Onsager's nonequilibrium thermodynamics, leads to dynamic "regression models" that should be useful for data analysis.

1.4 Cognition as "Language"

Atlan and Cohen (1998) argue above, in the context of the immune system, that cognitive function involves comparison of a perceived signal with an internal, learned or inherited picture of the world, and then choice of one response from a much larger repertoire of possible responses. That is, cognitive pattern recognition-and-response proceeds by an algorithmic combination of an incoming external sensory signal with an internal ongoing activity—incorporating the internalized picture of the world—and triggering an appropriate action based on a decision that the pattern of sensory activity requires a response.

Incoming sensory input is thus mixed in an unspecified but systematic manner with a pattern of internal ongoing activity to create a path of combined signals $x = (a_0, a_1, \ldots, a_n, \ldots)$. Each a_k thus represents some functional composition of the internal and the external. An application of this perspective to a standard neural network is given in Wallace (2005, p. 34).

This path is fed into a similarly unspecified, decision function, h, generating an output $h(x)$ that is an element of one of two disjoint sets B_0 and B_1 of possible system responses. Let

$$B_0 \equiv \{b_0, \ldots, b_k\},$$

$$B_1 \equiv \{b_{k+1}, \ldots, b_m\}.$$

Assume a graded response, supposing that if

$$h(x) \in B_0,$$

the pattern is not recognized, and if

$$h(x) \in B_1,$$

the pattern is recognized, and some action $b_j, k+1 \leq j \leq m$ takes place.

Formal interest focuses on paths x triggering pattern recognition-and-response: given a fixed initial state a_0, examine all possible subsequent paths x beginning with a_0 and leading to the event $h(x) \in B_1$. Thus $h(a_0, \ldots, a_j) \in B_0$ for all $0 \leq j < m$, but $h(a_0, \ldots, a_m) \in B_1$.

For each positive integer n, let $N(n)$ be the number of high probability paths of length n that begin with some particular a_0 and lead to the condition $h(x) \in B_1$. Call such paths "meaningful," assuming that $N(n)$ will be considerably less than the number of all possible paths of length n leading from a_0 to the condition $h(x) \in B_1$.

Note that identification of the "alphabet" of the states a_j, B_k may depend on the proper system "coarse graining" in the sense of symbolic dynamics (Beck and Schlogl 1995).

Combining algorithm, the form of the function h and the details of grammar and syntax are all unspecified in this model. The assumption permitting inference on necessary conditions constrained by the asymptotic limit theorems of information theory is that the finite limit

$$H \equiv \lim_{n\to\infty} \frac{\log[N(n)]}{n} \tag{1.1}$$

both exists and is independent of the path x.

Call such a pattern recognition-and-response cognitive process *ergodic*. Not all cognitive processes are likely to be ergodic, implying that H, if it indeed exists at all, is path dependent, although extension to nearly ergodic processes, in a certain sense, seems possible (e.g., Wallace 2005, pp. 31–32).

Invoking the spirit of the Shannon–McMillan Theorem, it is possible to define an adiabatically, piecewise stationary, ergodic information source $\mathbf{X}$ associated with stochastic variates X_j having joint and conditional probabilities $P(a_0, \ldots, a_n)$ and $P(a_n|a_0, \ldots, a_{n-1})$ such that appropriate joint and conditional Shannon uncertainties satisfy the classic relations (Cover and Thomas 2006)

$$H[\mathbf{X}] = \lim_{n\to\infty} \frac{\log[N(n)]}{n} = \lim_{n\to\infty} H(X_n|X_0, \ldots, X_{n-1}) = \lim_{n\to\infty} \frac{H(X_0, \ldots, X_n)}{n}. \tag{1.2}$$

This information source is defined as *dual* to the underlying ergodic cognitive process, in the sense of Wallace (2000, 2005, 2007).

"Adiabatic" means that, when the information source is parameterized according to some appropriate scheme, within continuous "pieces," changes in parameter values take place slowly enough so that the information source remains as close to stationary and ergodic as needed to make the fundamental limit theorems work. "Stationary" means that probabilities do not change in time, and "ergodic" (roughly) that cross-sectional means converge to long-time averages. Between "pieces" one invokes various kinds of phase change formalism, for example, renormalization theory in cases where a mean field approximation holds (Wallace 2005), or variants of random network theory where a mean number approximation is applied.

Extension of the theory to "nonergodic" information sources is possible, given some considerable mathematical overhead, e.g., Wallace (2005, pp. 31–32).

Recall that the Shannon uncertainties $H(\ldots)$ are cross-sectional law-of-large-numbers sums of the form $-\sum_k P_k \log[P_k]$, where the P_k constitute a probability distribution. See Cover and Thomas (2006), Ash (1990), or Khinchin (1957) for the standard details.

A formal equivalence class algebra can be constructed by choosing different origin points, a_0, and defining the equivalence of two states, a_m, a_n, by the existence of high probability meaningful paths connecting them to the same origin point. Disjoint partition by equivalence class, analogous to orbit equivalence classes for dynamical systems, defines the vertices of a network of cognitive dual languages. Each vertex then represents a different information source dual to a cognitive process. This is not a representation of a neural network as such, or of some circuit in silicon. It is, rather, an abstract set of "languages" dual to the set of cognitive biological processes.

This structure generates a groupoid, leading to complicated algebraic properties summarized in the Mathematical Appendix.

1.5 No Free Lunch

Given a set of biological cognitive modules that become linked to solve a problem—e.g., riding a bicycle in heavy traffic, followed by localized wound healing—the famous "no free lunch" theorem of Wolpert and MacReady (1995, 1997) illuminates the next step in the argument. As English (1996) states the matter,

> ... Wolpert and Macready... have established that there exists no generally superior [computational] function optimizer. There is no "free lunch" in the sense that an optimizer "pays" for superior performance on some functions with inferior performance on others... gains and losses balance precisely, and all optimizers have identical average performance... [That is] an optimizer has to "pay" for its superiority on one subset of functions with inferiority on the complementary subset...

Another way of stating this conundrum is to say that a computed solution is simply the product of the information processing of a problem, and, by a very famous argument, information can never be gained simply by processing. Thus a problem X is transmitted as a message by an information processing channel, Y, a computing device, and recoded as an answer. By the "tuning theorem" argument of the Mathematical Appendix, there will be a channel coding of Y which, when properly tuned, is *itself* most efficiently "transmitted," in a purely formal sense, by the problem—the "message" X. In general, then, the most efficient coding of the transmission channel, that is, the best algorithm turning a problem into a solution, will necessarily be highly problem-specific. Thus there can be no best algorithm for all sets of problems, although there will likely be an optimal algorithm for any given set.

Indeed, something much like this result is well known, using another description. Shannon (1959) wrote

> There is a curious and provocative duality between the properties of [an information] source with a distortion measure and those of a channel. This duality is enhanced if we consider channels in which there is a cost associated with the different letters... Solving this problem corresponds, in a sense, to finding a source that is right for the channel and the desired cost... In a somewhat dual way, evaluating the rate distortion function for a source... corresponds to finding a channel that is just right for the source and allowed distortion level.

From the no free lunch argument, it is clear that different challenges facing an entity must be met by different arrangements of cooperating "low level" cognitive modules. It is possible to make a very abstract picture of this phenomenon, not based on anatomy, but rather on the network of linkages between the information sources dual to the physiological and learned unconscious cognitive modules (UCM). That is, the remapped network of lower level cognitive modules is reexpressed in terms of the information sources dual to the UCM. Given two distinct problems classes (e.g., riding a bicycle vs. wound healing), there must be two different "wirings" of the information sources dual to the available physiological UCM, as in Fig. 1.2, with the network graph edges measured by the amount of information crosstalk between sets of nodes representing the dual information sources. A more formal treatment

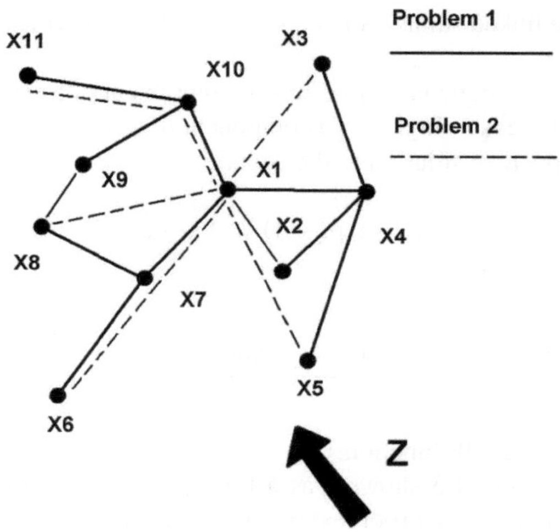

Fig. 1.2 By the no free lunch theorem, two markedly different problems facing an organism will be optimally solved by two different linkages of available lower level cognitive modules—characterized now by their dual information sources X_j—into different temporary networks of working structures, here represented by crosstalk among those sources rather than by the physiological UCM themselves. The embedding information source Z represents the influence of external signals whose effects can be at least formally accounted for by network information theory

of such coupling can be given in terms of network information theory (Cover and Thomas 2006), particularly incorporating the effects of embedding contexts, implied by the "external" information source Z—signals from the environment.

The possible expansion of a closely linked set of information sources dual to the UCM into a global broadcast—the occurrence of a kind of "spandrel"—depends, in this model, on the underlying network topology of the dual information sources and on the strength of the couplings between the individual components of that network.

For random networks the results are well known (Erdos and Renyi 1960). Following the review by Spenser (2010) closely (see, e.g., Boccaletti et al. 2006, for more detail), assume there are n network nodes and e edges connecting the nodes, distributed with uniform probability—no nonrandom clustering. Let $G[n, e]$ be the state when there are e edges. The central question is the typical behavior of $G[n, e]$ as e changes from 0 to $(n - 2)!/2$. The latter expression is the number of possible pair contacts in a population having n individuals. Another way to say this is to let $G(n, p)$ be the probability space over graphs on n vertices where each pair is adjacent with independent probability p. The behaviors of $G[n, e]$ and $G(n, p)$ where $e = p(n - 2)!/2$ are asymptotically the same.

For the simple random case, parameterize as $p = c/n$. The graph with $n/2$ edges then corresponds to $c = 1$. The essential finding is that the behavior of the random network has three sections:

1. If $c < 1$, all the linked subnetworks are very small, *and no global broadcast can take place*.
2. If $c = 1$, there is a single large interlinked component of a size $\approx n^{2/3}$.
3. If $c > 1$, then there is a single large component of size yn—a global broadcast—where y is the positive solution to the equation

$$\exp(-cy) = 1 - y\,. \tag{1.3}$$

Then

$$y = \frac{W(-c/\exp(c)) + c}{c}, \tag{1.4}$$

where W is the Lambert W function.

The solid line in Fig. 1.3 shows y as a function of c, representing the fraction of network nodes that are incorporated into the interlinked giant component—a de-facto global broadcast for interacting UCM. To the left of $c = 1$ there is no giant component, and large-scale cognitive process is not possible.

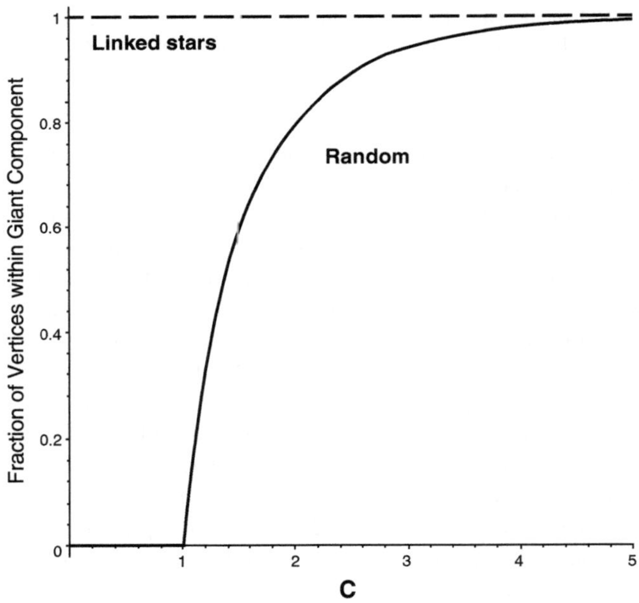

Fig. 1.3 Fraction of network nodes in the giant component as a function of the crosstalk coupling parameter c. The *solid line* represents a random graph, the *dotted line* a star-of-stars-of-stars network in which all nodes are interconnected, showing that the dynamics of giant component emergence are highly dependent on an underlying network topology that, for UCM, may itself be tunable. For the random graph, a strength of $c < 1$ precludes emergence of a larger-scale "global" broadcast

The dotted line, however, represents the fraction of nodes in the giant component for a highly nonrandom network, a star-of-stars-of-stars (SoS) in which every node is directly or indirectly connected with every other one. For such a topology there is no threshold, only a single giant component, showing that the emergence of a giant component in a network of information sources dual to the UCM is dependent on a network topology that may itself be tunable. A generalization of this result follows from an index theorem argument below.

1.6 Multiple Broadcasts, Punctuated Detection

The random network development above is predicated on there being a variable average number of fixed-strength linkages between components. Clearly, the mutual information measure of crosstalk is not inherently fixed, but can continuously vary in magnitude. This suggests a parameterized renormalization. In essence, the modular network structure linked by mutual information interactions has a topology depending on the degree of interaction of interest.

Define an interaction parameter ω, a real positive number, and look at geometric structures defined in terms of linkages set to zero if mutual information is less than, and "renormalized" to unity if greater than, ω. Any given ω will define a regime of giant components of network elements linked by mutual information greater than or equal to it.

Now invert the argument: A given topology for the giant component will, in turn, define some critical value, ω_C, so that network elements interacting by mutual information less than that value will be unable to participate, i.e., will be locked out and not be "consciously" perceived. Thus ω is a tunable, syntactically dependent, detection limit that depends critically on the instantaneous topology of the giant component of linked cognitive modules defining the global broadcast. That topology is, fundamentally, the basic tunable syntactic filter across the underlying modular structure, and variation in ω is the only one aspect of a much more general topological shift. Further analysis can be given in terms of a topological rate distortion manifold (Glazebrook and Wallace 2009a,b).

There is considerable empirical evidence from fMRI brain imaging and many other experiments to show that individual animal consciousness—restricted by necessity of a time constant near 100 ms—involves a single, shifting and tunable, global broadcast, a matter leading to the phenomenon of inattentional blindness. Multiple cognitive submodules within systems not constrained to the 100 ms time range, for example, institutions—individuals, departments, formal and informal workgroups—by contrast, can do more than one thing, and indeed, are usually required to multitask. Clearly, then, multiple global broadcasts—indexed by a set $\Omega = \{\omega_1, \ldots, \omega_n\}$—lessen the probability of inattentional blindness, if there is time to support them, but do not eliminate it, and introduce critical failure modes related to the degradation of information transmitted between global broadcasts.

Thus Ω represents a set of crosstalk information measures between cognitive submodules, each associated with its own tunable giant component having its own special topology.

Again, although animal consciousness, with its 100 ms time constant, seems restricted to a single tunable global broadcast, it is clear that slower physiological global broadcast analogs would permit individual subsystems, or localized sets of such subsystems, to engage in more than one global broadcast at a time, to multitask, in the same sense that workgroups within an institution will usually be given more than one task at a time. Thus the immune system can be expected to simultaneously engage in wound healing, attack on invading microorganisms, neuroimmuno dialog, and routine tissue maintenance tasks (Cohen 2000).

1.7 Metabolic Constraints

The information sources dual to the linked unconscious cognitive modules represented in Fig. 1.2 are not independent, but are correlated, so that a joint information source can be defined having the properties

$$H(X_1, \ldots, X_n) \leq \sum_{j=1}^{n} H(X_j) \tag{1.5}$$

This is the information chain rule (e.g., Cover and Thomas 2006), and has profound implications. Feynman (2000) describes in great detail how information and free energy have an inherent duality. Feynman, in fact, defines information precisely as the free energy needed to erase a message. The argument is surprisingly direct (e.g., Bennett 1988), and for very simple systems it is easy to design a small (idealized) machine that turns the information within a message directly into usable work—free energy. Information is a form of free energy and the construction and transmission of information within living things—the physical instantiation of information—consumes metabolic free energy, with inevitable and considerable losses via the second law of thermodynamics.

Suppose an intensity of metabolic free energy is associated with each information source $H(X,Y), H(X), H(Y)$, e.g., rates $M_{X,Y}, M_X, M_Y$.

Although information is a form of free energy, in the sense of Feynman (2000) and Bennett (1988), there is a massive entropic loss in its physical expression, so that the probability distribution of a source uncertainty H can be written as

$$P[H] = \frac{\exp[-H/\kappa M]}{\int \exp[-H/\kappa M] dH} \tag{1.6}$$

assuming κ is very small.

To first order,

$$\hat{H} = \int HP[H]dH \approx \kappa M \tag{1.7}$$

and, using Eq. (1.5),

$$\begin{aligned}\hat{H}(X,Y) &\leq \hat{H}(X) + \hat{H}(Y)\\ M_{X,Y} &\leq M_X + M_Y\end{aligned} \tag{1.8}$$

Thus, as a consequence of the information chain rule, allowing crosstalk consumes a lower rate of metabolic free energy than isolating information sources. A more detailed calculation is given in Sect. 11.3 of the Mathematical Appendix. That is, in general, it takes more metabolic free energy to isolate a set of cognitive phenomena than it does to allow them to engage in crosstalk, a signal interaction that, under typical electrical engineering circumstances, grows as the inverse square of the separation between circuits. This is a well-known problem in electrical engineering that can consume considerable attention and other resources for proper address.

The global broadcast mechanisms of consciousness and its slower physiological generalizations make an arch of this spandrel, using the lowered metabolic free energy requirement of crosstalk interaction between low level cognitive modules as the springboard for launching (sometimes) rapid, tunable, more highly correlated, multiple global broadcasts that link those modules to solve problems.

1.8 Environmental Signals

Lower level cognitive modules operate within larger, highly structured, environmental signals and other constraints whose regularities may also have a recognizable grammar and syntax, represented in Fig. 1.2 by an embedding information source Z. Under such a circumstance the splitting criterion for three jointly typical sequences is given by the classic relation of network information theory (Cover and Thomas 2006, Theorem 15.2.3)

$$I(X_1, X_2|Z) = H(Z) + H(X_1|Z) + H(X_2|Z) - H(X_1, X_2, Z), \tag{1.9}$$

that generalizes as

$$I(X_1, \ldots, X_n|Z) = H(Z) + \sum_{j=1}^{n} H(X_j|Z) - H(X_1, \ldots, X_n, Z) \tag{1.10}$$

More complicated multivariate typical sequences are treated much the same (e.g., El Gamal and Kim 2010, pp. 2–26). Given a basic set of interacting information sources $(X_1, \ldots, X_k)$ that one partitions into two ordered sets $X(\mathcal{J})$ and $X(\mathcal{J}')$, then the splitting criterion becomes $H[X(\mathcal{J}|\mathcal{J}')]$. Extension to a greater number of ordered sets is straightforward.

Then the joint splitting criterion—I, H above—however it may be expressed as a composite of the underlying information sources and their interactions, satisfies a relation like the first expression in Eq. (1.2), where $N(n)$ is the number of high probability jointly typical paths of length n, and the theory carries through, now incorporating the effects of external signals as the information source Z.

1.9 Dynamic "Regression Models"

Given the splitting criteria $I(X_1, \ldots, X_n|Z)$ or $H[X(\mathcal{J}|\mathcal{J}')]$ as above, the essential point is that these are the limit, for large n, of the expression $\log[N(n)]/n$, where $N(n)$ is the number of jointly typical high probability paths of the interacting information sources of length n. Again, as Feynman (2000) argues at great length, information is simply another form of free energy, and its dynamics can be expressed using a formalism similar to Onsager's nonequilibrium thermodynamics (de Groot and Mazur 1984). This is particularly apt in view of the enormous levels of free energy needed to physically instantiate information transmission.

First, the physical model. Let $F(K)$ be the free energy density of a physical system, K the normalized temperature, V the volume, and $Z(K, V)$ the *partition function* defined from the Hamiltonian characterizing energy states E_i. Then

$$Z(V, K) \equiv \sum_i \exp[-E_i(V)/K] \equiv \exp[F/K] \tag{1.11}$$

so that

$$F(K) = \lim_{V \to \infty} -K \frac{\log[Z(V, K))}{V} \equiv \frac{\log[\hat{Z}(K, V)]}{V}.$$

If a nonequilibrium physical system is parameterized by a set of variables $\{K_i\}$, then the *empirical Onsager equations* are defined in terms of the gradient of the entropy $S \equiv F - \sum_j K_j dF/dK_j$—the Legendre transform of F—as

$$dK_j/dt = \sum_i L_{i,j} \partial S/\partial K_i \tag{1.12}$$

where the $L_{i,j}$ are empirical constants. For a physical system having microreversibility, $L_{i,j} = L_{j,i}$. For an information source where, for example, "the" has a much different probability than "eht," no such microreversibility is possible, and no "reciprocity relations" can apply.

For stochastic systems this generalizes to the set of stochastic differential equations

$$dK_t^j = \sum_i [L_{j,i}(t, \ldots \partial S/\partial K^i \ldots)dt + \sigma_{j,i}(t, \ldots \partial S/\partial K^i)dB_t^i]$$
$$= L(t, K^1, \ldots, K^n)dt + \sum_i \sigma(t, K^1, \ldots, K^n)dB_t^i \quad (1.13)$$

where terms have been collected and expressed in the driving parameters. The dB_t^i represent different kinds of "noise" whose characteristics are usually expressed by their quadratic variation. See any standard text for definitions, examples, and details (e.g., Protter 1990).

For the splitting criteria $I(X_1, \ldots, X_n|Z)$ or $H[X(\mathcal{J}|\mathcal{J}')]$, the role of information as a form of free energy and the corresponding limit in $\log[N(n)]/n$ make it possible to define entropy-analogs as

$$S \equiv I(\ldots K^i \ldots) - \sum_j K^j \partial I/\partial K^j,$$
$$S \equiv H[X(\mathcal{J}|\mathcal{J}')] - \sum_j K^j \partial H[X(\mathcal{J}|\mathcal{J}')]/\partial K^j,$$
$$S \propto M_{\mathcal{J}|\mathcal{J}'} - \sum_j K^j \partial M_{\mathcal{J}|\mathcal{J}'}/\partial K^j \quad (1.14)$$

where the last relation invokes the embedding metabolic free energies that instantiate the actual mechanisms by which information is transmitted.

The basic dynamic "regression equations" for the system of Figs. 1.2 and 1.3, driven by a set of external "sensory" and other, internal, signal parameters $\mathbf{K} = (K^1, \ldots, K^n)$ that may be measured by the information source uncertainty of other information sources, are then precisely the set of Eq. (1.13) above.

That is, the underlying picture becomes reversed, and the actual driving metabolic free energy measures M_X are now seen as indexed by the source uncertainties $H[X]$. The different M_X become each others' embedding environments in an analog to coevolutionary dynamics.

Several features emerge directly from invoking such a coevolutionary approach.

1. Setting the expectation of Eq. (1.13) equal to zero and solving for stationary points gives attractor states since the noise terms preclude unstable equilibria.
2. This system may converge to limit cycle or pseudorandom "strange attractor" behaviors in which the system seems to chase its tail endlessly within a limited venue—a kind of "Red Queen" pathology.
3. What is converged to in both cases is not a simple state or limit cycle of states. Rather it is an equivalence class, or set of them, of highly dynamic information

sources coupled by mutual interaction through crosstalk. Thus "stability" in this structure represents particular patterns of ongoing dynamics rather than some identifiable static configuration.

This represents a highly recursive phenomenological set of stochastic differential equations (Zhu et al. 2007), but operates in a dynamic rather than static manner. The objects of this dynamical system are equivalence classes of information sources, rather than simple "stationary states" of a dynamical or reactive chemical system. The necessary conditions of the asymptotic limit theorems of communication theory have beaten the mathematical thicket back one layer.

Third, as Champagnet et al. (2006) note, shifts between the quasi-equilibria of a coevolutionary system can be addressed by the large deviations formalism. The issue of dynamics drifting away from trajectories predicted by the canonical equation can be investigated by considering the asymptotic of the probability of "rare events" for the sample paths of the diffusion.

"Rare events" are the diffusion paths drifting far away from the direct solutions of the canonical equation. The probability of such rare events is governed by a large deviation principle: when a critical parameter (designated ϵ) goes to zero, the probability that the sample path of the diffusion is close to a given rare path ϕ decreases exponentially to 0 with rate $\mathcal{I}(\phi)$, where the "rate function" $\mathcal{I}$ can be expressed in terms of the parameters of the diffusion.

This result can be used to study long-time behavior of the diffusion process when there are multiple attractive singularities. Under proper conditions the most likely path followed by the diffusion when exiting a basin of attraction is the one minimizing the rate function $\mathcal{I}$ over all the appropriate trajectories. The time needed to exit the basin is of the order $\exp(\mathcal{V}/\epsilon)$ where $\mathcal{V}$ is a quasi-potential representing the minimum of the rate function $\mathcal{I}$ over all possible trajectories.

An essential fact of large deviations theory is that the rate function $\mathcal{I}$ that Champagnat et al. invoke can be expressed as a kind of formal entropy measure, that is, having the canonical form

$$\mathcal{I} = -\sum_j P_j \log(P_j) \tag{1.15}$$

for some probability distribution. This result goes under a number of names: Sanov's Theorem, Cramer's Theorem, the Gartner–Ellis Theorem, the Shannon–McMillan Theorem, and so forth (Dembo and Zeitouni 1998).

These arguments are very much in the direction of Eq. (1.13), now seen as subject to internally driven large deviations that can themselves be described as information sources, providing $K = f(\mathcal{I})$-parameters that can trigger punctuated shifts between quasi-stable modes. Thus both external signals, characterized by the information source Z, and internal "ruminations," characterized by an information source $\mathcal{I}$, can provide K-parameters that serve to drive the system to different quasi-equilibrium "conscious attention states" in a highly punctuated manner, if they are of sufficient magnitude to overcome the topological renormalization ω-constraints described in Sect. 1.6.

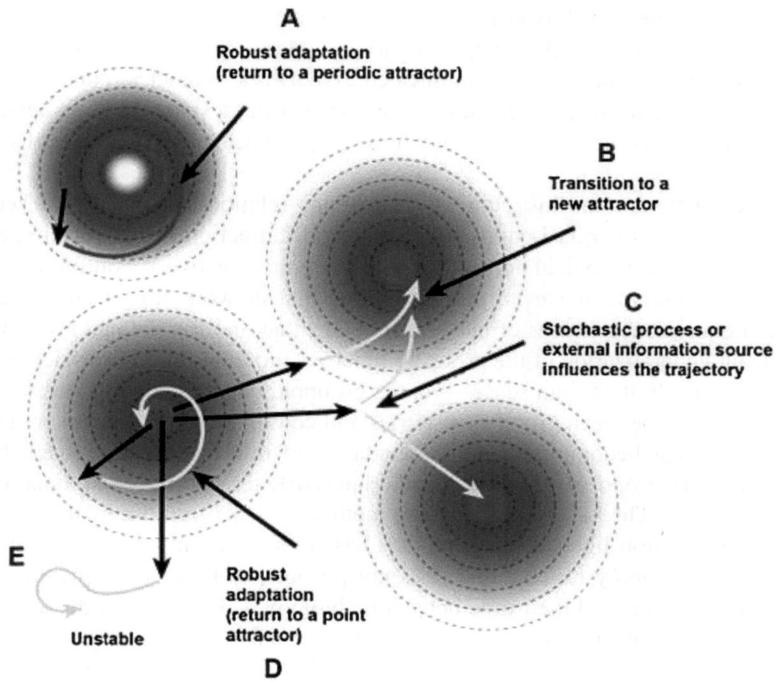

Fig. 1.4 Adapted from Fig. 1 of Kitano (2004). Using a two-dimensional schematic, different basins of attraction in parameter space show system response to perturbation. (*A*) Return to a periodic (or chaotic) attractor. (*B*) Transition to a new attractor. (*C*) Stochastic process or external information source $\mathcal{I}$ influences the trajectory. (*D*) Return to a point attractor. (*E*) Unstable random walk to another basin of attraction. Compare with Fig. 1.2 which represents the different network topologies of the dual information sources that underlie the parameterization

A schematic of these ideas has become common currency in systems biology, and Fig. 1.4, adapted from Fig. 1 of Kitano (2004), provides a caricature, including several possible modes. Taking a two-dimensional parameterization, so that there are two K_j, different basins of attraction in parameter space show system response to perturbation: (A) Return to a periodic (or chaotic) attractor. (B) Transition to a new attractor. (C) Stochastic process or external information source $\mathcal{I}$ influences the trajectory. (D) Return to a point attractor. (E) An unstable random walk to another basin of attraction. This should be compared with Fig. 1.2 that represents two of the different network topologies of the dual information sources behind the parameterization.

Figure 1.4 then presents, in parameter space, several of the different "no free lunch" arrangements of unconscious cognitive modules, as in Fig. 1.2, that unite to address problems facing the organism.

More generally, however, following the topological arguments of Sect. 1.6, setting the expectation of Eq. (1.13) to zero generates an index theorem (Hazewinkel

2002), in the sense of Atiyah and Singer (1963). Such an object relates analytic results—the solutions to the equations—to an underlying set of topological structures that are eigenmodes of a complicated Ω-network geometric operator whose spectrum represents the possible multiple global broadcast states of the system. This structure and its dynamics do not really have simple mechanical or electrical system analogs.

Index theorems, in this context, instantiate relations between "conserved" quantities—here, the quasi-equilibria of basins of attraction in parameter space—and underlying topological form—here, the cognitive network conformations of Fig. 1.2. Section 1.4, however, described how that network was itself defined in terms of equivalence classes of meaningful paths that, in turn, defined groupoids, a significant generalization of the group symmetries more familiar to physicists.

The approach, then, in a sense—via the groupoid construction—generalizes the famous relation between group symmetries and conservation laws uncovered by E. Noether that has become the central foundation of modern physics (Byers 1999). Thus this work proposes a kind of Noetherian (NER-terian) statistical dynamics of the living state. The saving grace of the method is that it represents the fitting of dynamic regression-like statistical models based on the asymptotic limit theorems of information theory to data, and does not presume to be a "real" picture of the underlying systems and their time behaviors: biology is not relativity theory.

As with simple fitted regression equations, actual scientific inference is done most often by comparing the same systems under different, and different systems under the same, conditions. Statistics is not science, and one can easily imagine the necessity of "nonparametric" or "non-Noetherian" models.

1.10 Phase Transition Approaches

Basic Ideas

Given sufficient available metabolic free energy, it is possible to refine the topological "renormalization" arguments of Sect. 1.6 in terms of the joint uncertainty measure driven by changes in the coupling parameter ω. Joint dynamic trajectories are assumed constrained by crosstalk, as indexed by ω, so that the probability distribution density function of a joint information source representing linked cognitive submodules is given, in a first approximation, by the standard expression for the Gibbs distribution

$$P[H(X_1, \ldots, X_n)] = \frac{\exp[-H/\kappa\omega]}{\int \exp[-H/\kappa\omega]dH}, \tag{1.16}$$

where κ is a scaling constant. Roughly, increasing the crosstalk measure ω permits higher "symmetries" by allowing more interaction.

The Gibbs distribution may not work well for systems whose dynamics are thermodynamically open, and it is possible to generalize the treatment, using an adiabatic approximation in which the dynamics remain "close enough" to a form in which the mathematical theory can work, adapting standard phase transition formalism for shifts between adiabatic realms. In particular, rather than using exponential terms, one might well use any functional form $g(H, \omega)$ such that the integral over H converges.

A partition function-analog can be defined as

$$Z(\kappa\omega) = \int \exp[-H/\kappa\omega]dH \tag{1.17}$$

Now define a "free energy," F, over the full set of possible dynamic trajectories as constrained by ω, as

$$\exp[-F/\kappa\omega] \equiv \int \exp[-H/\kappa\omega]dH, \tag{1.18}$$

so that

$$F(\kappa\omega) = -\kappa\omega \log[Z(\kappa\omega)]\,. \tag{1.19}$$

This is to be taken as a Morse Function, in the sense of the Mathematical Appendix. Other—essentially similar—Morse Functions may perhaps be defined on this system, having a more "natural" interpretation from information theory.

Argument is now by abduction from statistical physics (Landau and Lifshitz 2007; Pettini 2007). The Morse Function F is seen as constrained by the crosstalk linkage parameter ω in a manner that allows application of Landau's theory of punctuated phase transition in terms of groupoid, rather than group, symmetries.

Recall Landau's perspective on phase transition (Pettini 2007). The essence of his insight was that certain physical phase transitions took place in the context of a significant symmetry change, with one phase being more symmetric than the other. A symmetry is lost in the transition, i.e., spontaneous symmetry breaking. The greatest possible set of symmetries being that of the Hamiltonian describing the energy states. Usually, states accessible at lower temperatures will lack the symmetries available at higher temperatures, so that the lower temperature state is less symmetric, and transitions can be highly punctuated.

Here, dynamic process is characterized in terms of groupoid, rather than group, symmetries, and the argument by abduction is essentially similar: Increasing crosstalk—rising ω—will allow richer interactions between the interacting information sources, and will do so in a highly punctuated manner.

Kadanoff Theory

Given F as a free energy analog, a mathematical treatment of transitions between adiabatic realms is of interest. Define a characteristic "length," r, on the network of interacting information sources, as more fully described below. It is then possible to apply renormalization symmetries in terms of the "clumping" transformation, so that, for clumps of size R, in an external "field" of strength J (that can be set to 0 in the limit), in the usual manner (e.g., Wilson 1971),

$$F[\omega(R), J(R)] = f(R)F[\omega(1), J(1)]$$
$$\chi(\omega(R), J(R)) = \frac{\chi(\omega(1), J(1))}{R} \tag{1.20}$$

where χ is a characteristic correlation length.

As will be described below, many "biological" renormalizations, $f(R)$, are possible that lead to a number of quite different universality classes for phase transition.

In order to define the metric r, impose a topology on the system of interacting information sources, so that, near a particular "language" A defining some uncertainty measure H there is (in an appropriate sense) an open set U of closely similar languages $\hat{A}$, such that $A, \hat{A} \subset U$.

Since the information sources are "similar," for all pairs of languages $A, \hat{A}$ in U, it is possible to:

1. Create an embedding alphabet which includes all symbols allowed to both of them.
2. Define an information-theoretic distortion measure in that extended, joint alphabet between any high probability (grammatical and syntactical) paths in A and $\hat{A}$, which we write as $d(Ax, \hat{A}x)$ (Cover and Thomas 2006). Note that these languages do not interact, in this approximation.
3. Define a metric on U, for example,

$$r(A, \hat{A}) = |\lim \frac{\int_{A,\hat{A}} d(Ax, \hat{A}x)}{\int_{A,A} d(Ax, A\hat{x})} - 1|, \tag{1.21}$$

using an appropriate integration limit argument over the high probability paths. Note that the integration in the denominator is over different paths within A itself, while in the numerator it is between different paths in A and $\hat{A}$. Consideration suggests r is indeed a formal metric.

Other Versions of the Kadanoff Model

Clearly, other ways of constructing a metric seem possible, as are other ways of renormalizing. For example, Wallace (2005) uses a more direct argument in which the richness of the joint uncertainty across the network of interacting cognitive modules itself grows according to something like Eq. (1.20). While Wilson's (1971) calculation necessarily had a volume functional dependence, $f(R) = R^3$, source uncertainty is likely to "top out" and not increase without limit, or at least not so rapidly, and Wallace (2005) explores the influence of different forms of $f(R)$: $R^\delta, m\log(R) + 1, \exp[m(R-1)/R]$. Surprisingly, a variety of functional forms for $f(R)$ carries through using this technique.

1.11 The Rate Distortion Approach

Some Formalism

A different perspective regarding cognitive global broadcasts emerges through an index theorem attack based on an Onsager-like nonequilibrium treatment of the disjunction between intent and impact of cognitive subprocesses, in the context of "perturbations" that may range from incoming sensory signals to pathogens, developing tumors, and the like. This is done via the Rate Distortion Theorem, leading to a conceptual simplification of previous arguments.

Many real-time biological problems are inherently rate distortion problems, in the same sense that the retina is a tool for projection of complex visual stimuli down onto a "simpler" neural substrate, and it is possible to reformulate the underlying theory from that perspective. The implementation of a complex cognitive structure, say a sequence of control orders generated by some regulatory dual information source Y, having output $y^n = y_1, y_2, \ldots$ is "digitized" in terms of the observed behavior of the regulated system, say the sequence $b^n = b_1, b_2, \ldots$. The b_i are thus what happens in real time, the actual impact of the cognitive structure on its embedding environment. Assume each b^n is then deterministically retranslated back into a reproduction of the original control signal, $b^n \to \hat{y}^n = \hat{y}_1, \hat{y}_2, \ldots$.

Define a distortion measure $d(y, \hat{y})$ that compares the original to the retranslated path. See Cover and Thomas (2006) for example. Suppose that with each path y^n and b^n-path retranslation into the y-language, denoted $\hat{y}^n$, there are associated individual, joint, and conditional probability distributions $p(y^n), p(\hat{y}^n), p(y^n|\hat{y}^n)$.

The average distortion is defined as

$$D \equiv \sum_{y^n} p(y^n) d(y^n, \hat{y}^n) . \tag{1.22}$$

It is possible, using the distributions given above, to define the information transmitted from the incoming Y to the outgoing $\hat{Y}$ process using the Shannon source uncertainty of the strings:

$$I(Y, \hat{Y}) \equiv H(Y) - H(Y|\hat{Y}) = H(Y) + H(\hat{Y}) - H(Y, \hat{Y}).$$

If there is no uncertainty in Y, given the retranslation $\hat{Y}$, then no information is lost, and the regulated system is perfectly under control.

In general, this will not be true.

The information Rate Distortion Function $R(D)$ for a source Y with a distortion measure $d(y, \hat{y})$ is defined as

$$R(D) = \min_{p(y,\hat{y});\sum_{(y,\hat{y})} p(y)p(y|\hat{y})d(y,\hat{y}) \leq D} I(Y, \hat{Y}) \tag{1.23}$$

Cover and Thomas (2006) provide more detail.

The minimization is over all conditional distributions $p(y|\hat{y})$ for which the joint distribution $p(y, \hat{y}) = p(y)p(y|\hat{y})$ satisfies the average distortion constraint (i.e., average distortion $\leq D$).

The Rate Distortion Theorem states that $R(D)$ is the minimum necessary rate of information transmission—essentially minimum channel capacity—that ensures the transmission does not exceed average distortion D (Cover and Thomas 2006). The Rate Distortion Function has been calculated for a number of systems, using Lagrange multiplier methods. Cover and Thomas (2006) show that $R(D)$ is necessarily a decreasing convex function of D, that is, always a reverse J-shaped curve. This is a critical observation, since convexity is an exceptionally powerful mathematical condition (Ellis 1985; Rockafellar 1970).

Recall, now, the classic relation between information source uncertainty and channel capacity. First,

$$H[\mathbf{X}] \leq C, \tag{1.24}$$

where H is the uncertainty of the source X and C the channel capacity. Recall that C is defined according to the relation

$$C \equiv \max_{P(X)} I(X|Y), \tag{1.25}$$

where $P(X)$ is the probability distribution of the message chosen so as to maximize the rate of information transmission along a channel Y.

A Simple Model

The Rate Distortion Function places limits on information source uncertainty. Thus distortion measures can drive information system dynamics. That is, the Rate Distortion Function itself has a homological relation to free energy density.

The motivation for this approach is the observation that a Gaussian channel with noise variance σ^2 and zero mean has a Rate Distortion Function $R(D) = 1/2 \log[\sigma^2/D]$ using the squared distortion measure. Defining a "Rate Distortion entropy" as the Legendre transform of the Rate Distortion Function

$$S_R \equiv R(D) - DdR(D)/dD = 1/2 \log[\sigma^2/D] + 1/2,$$

the simplest possible nonequilibrium Onsager equation (de Groot and Mazur 1984) is

$$dD/dt = -\mu dS_R/dD = \mu/2D,$$

where t is the time and μ is a diffusion coefficient. By inspection,

$$D(t) = \sqrt{\mu t}$$

very precisely the solution to the diffusion equation.

Some thought will suggest this correspondence reduction is of singular importance, and it is now possible to argue upward from it in both scale and complexity.

Suppose the relation between system challenge and system response—the manner in which physiological activities of the cognitive systems of interest more-or-less accurately reflect what is called for by environmental conditions—is characterized by another Gaussian channel. Again defining a rate distortion entropy as $S_R = R(D) - DdR/dD$ permits definition of a more complicated nonequilibrium Onsager equation in the presence of an incoming system perturbation δT as

$$dD/dt = -\mu dS_R/dD - 1/\delta T = \frac{\mu}{2D} - 1/\delta T \tag{1.26}$$

having the equilibrium solution, where $dD/dt = 0$,

$$D_{\text{equilib}} = \frac{\mu}{2}\delta T \tag{1.27}$$

The perturbation might be a sensory or regulatory signal, an incoming pathogen, a growing tumor, and so on.

For this simplistic model, the distortion between physiological need and actual physiological response is proportional to the perturbation. In reality, the final distortion measure D is the consequence of a vast array of internal processes that all contribute to it and that, individually, must all be optimized for the organism to survive. That is, overall distortion between total need and total response cannot be constrained by allowing critical subsystems to overload beyond survivable limits: Too high a blood pressure spike in response to a stress spike can be fatal.

A Less Simple Model

Consider individual responses of the interacting cognitive physiological systems, for the moment, to be a set of m independent but not identically distributed normal random variates having zero mean and variance $\sigma_i^2, i = 1 \ldots m$. Following the argument of Sect. 10.3.3 of Cover and Thomas (2006), assume a fixed channel capacity R available with which to represent this random vector. How should we allot signal to the various components to minimize the total distortion D? A brief argument shows it necessary to optimize

$$R(D) = \min_{\sum D_i = D} \sum_{i=1}^{m} \max\{1/2 \log[\sigma_i^2/D_i], 0\} \tag{1.28}$$

subject to the inequality restraint $\sum_i D_i \leq D$.

Using the Kuhn–Tucker generalization of the Lagrange multiplier method necessary under inequality conditions (e.g., Nocedal and Wright 1999) gives

$$R(D) = \sum_{i=1}^{m} 1/2 \log[\sigma_i^2/D_i], \tag{1.29}$$

where $D_i = \lambda$ if $\lambda < \sigma_i^2$ or $D = \sigma_i^2$ if $\lambda \geq \sigma_i^2$, and λ is chosen so that $\sum_i D_i = D$.

Thus, even under conditions of "independence," there is a complex "reverse water-filling" relation for Gaussian variables.

In the real world, the different subcomponents will engage in complicated crosstalk.

Assume m different subsystems that are not independent. Define a Rate Distortion function $R(D_1, \ldots, D_m) = R(\mathbf{D})$ and an associated Legendre transform "rate distortion entropy" S_R having the Onsager-like form

$$S_R = R(\mathbf{D}) - \sum_j D_j \partial S_R / \partial D_j \tag{1.30}$$

The most direct generalization of Eq. (1.28) is

$$dD_i/dt = -\sum_j L_{i,j} \partial S_R / \partial D_j - 1/\delta T_i \tag{1.31}$$

At equilibrium, all $dD_i/dt \equiv 0$, so that it becomes necessary to minimize *each* D_i under the joint constraints

$$\left[\sum_j L_{i,j}\partial S_R/\partial D_j\right] + 1/\delta T_i = 0,$$
$$D_i \leq D_i^{\max}\ \forall i \tag{1.32}$$

remembering that $R(\mathbf{D})$ must be a convex function.

The $D_i^{\max}$ represent limits on both internal and external distortion measures as needed for survival.

This is a complicated problem in Kuhn–Tucker optimization for which the exact form of the crosstalk-dominated $R(\mathbf{D})$ is quite unknown, in the context that even the independent Gaussian channel example involves constraints of mutual influence via reverse water-filling. In sum, changing a single perturbation δT_i will inevitably reverberate across the entire system, necessarily affecting—sometimes markedly—each distortion measure that characterizes the difference between needed and observed physiological subsystem response to challenge.

Most importantly, there may, in fact, be no general solution having $D_i \leq D_i^{\max} \forall\ i$, that is, no possible Pareto surface defining the limits of optimality. Such failure of solution is precisely the punctuated accession to detection of that perturbation. The setting of the variates $D_i^{\max}$ represents the tuning of the system of interacting cognitive modules. More subtle tuning, in the presence of noise, leads to the final model.

The Index Theorem Attack

The model of Eq. (1.31) admits unstable equilibria. Their elimination, and the imposition of more general tuning criteria, can be met by an appropriate system of stochastic Onsager differential equations, having the form

$$dD_t^i = [L_i(t, \mathbf{D}) - 1/f_i(\mathbf{T})]dt + \sigma_i(t, \mathbf{D})dB_t^i,$$
$$D_i \leq D_i^{\max}\ \forall i \tag{1.33}$$

where, again, the dB_t^i represent noise terms having characteristic quadratic variations (Protter 1990) and the $f_i(\mathbf{T})$ are monotonic increasing functions of the perturbation vector $\mathbf{T} = (\delta T_1, \ldots, \delta T_n)$. As above, noise precludes unstable equilibria, and is thus quite as important as crosstalk, interpreted as the exaptation of both noise and crosstalk into physiological global broadcasts.

Thus both the vectors $\mathbf{D} \equiv (D_1, \ldots, D_n)$ and $\mathbf{T}$ provide tuning criteria under the stochastic Kuhn–Tucker optimization conditions that would generalize Eq. (1.32).

That is, **D** establishes thresholds for perturbation detection, and the $f_i(\mathbf{T})$ tune the sensitivity of the system across the perturbation vector **T**, determining what will be "looked for" under nominal circumstances. Amplified perturbations that resonate across the system, and cause some D_i to exceed its D_i^{max}, enter the "theater of generalized consciousness" for the particular set of linked, crosstalking cognitive modules being brought into collaboration.

Note that Eq. (1.33) can be more simply expressed as

$$dD_t^i = \mathcal{L}_i(t, \mathbf{D}, \mathbf{T})dt + \sigma_i(t, \mathbf{D})dB_t^i,$$
$$D_i \leq D_i^{max} \; \forall i \tag{1.34}$$

Then the stochastic Kuhn–Tucker optimization is across the system of equations

$$[\mathcal{L}_i(t, \mathbf{D}, \mathbf{T})dt + \sigma_i(t, \mathbf{D})dB_t^i] = 0,$$
$$D_i \leq D_i^{max} \; \forall i \tag{1.35}$$

at a fixed perturbation setting **T**.

By the network linkages inherent in the functions $\mathcal{L}_j(t, \mathbf{D}, \mathbf{T})$, a perturbation δT_k can influence more than just the distortion measure D_k. That is, a perturbation δT_k that does not trigger a particular $D_k > D_k^{max}$ may still resonate across the system's crosstalk connections, violating an apparently distant D_j constraint, $j \neq k$.

Equation (1.35) represents, then, a generalized index theorem, in the sense discussed above, in that—underlying the analytic conditions—there are particular topologies of interconnected dual information sources linked by crosstalk, as described earlier. The details, however, can become mathematically complicated (e.g., Glazebrook and Wallace 2009a,b).

This argument provides another approach, via necessary conditions imposed by the asymptotic limits of information theory, to empirical models for a broad spectrum of global broadcast phenomena that recruit individual cognitive modules into shifting cooperative arrays that have both tunable detection thresholds for perturbation and tunable sensitivities to perturbation. These dynamic rate distortion models are, again, analogous to empirical regression models based on the Central Limit Theorem. When optimization by an "unconscious" cognitive system fails—there is no Pareto surface—the signal is propelled, in a punctuated manner, into detection by the "spotlight" that represents accession to attention from a global broadcast of linked subsystems needed to address a problem too complex for any single submodule.

1.12 Discussion and Conclusions

Sections 1.9–1.11 explore related statistical models of cognitive global broadcasts, slightly different views of the same elephant, as it were. While useful for data analysis—analogous to the varieties of regression models—they also provide examples of Dretske's imperative: the necessary conditions imposed by the asymptotic limit theorems of communication theory constrain cognitive process at all scales, levels of organization, and modes of distribution.

Information is a form of free energy instantiated by physical processes that themselves consume free energy, permitting adaptation of empirical approaches from nonequilibrium thermodynamics and statistical mechanics to cognitive phenomena. There are, however, restrictions imposed by the local irreversibility of information sources. Embedding high-level neural global broadcasts within a nested hierarchy of cognitive and other sources of impinging information evades the logical fallacy of attributing to "the brain" and the broad spectrum of functions that can only be embodied by the full individual-in-context.

The mereological fallacy is, then, fundamentally a matter of decontextualization. Bennett and Hacker (2003) use the lever of that fallacy to banish a philosophical *ignis fatuus* long afflicting consciousness studies that ranges across "the redness of red," "qualia," the "hard problem," "what is it like to be a bat," and so on. Further, Baars' global broadcast model makes clear that consciousness is very much an expression of classical physics and biochemistry that does not require reification to some new quality like mass or charge, or invocation of room-temperature quantum computing (Tegmark 2000).

There remains, however, an implicit and no less debilitating decontextualization that assumes individual consciousness is not merely different from other evolutionary adaptations, but exceptionally so. This belief has deep cultural roots, of concern, perhaps, to anthropologists or social scientists engaged in the study of religion. The biological or physical science context for the study of consciousness, however, must explicitly include the evolutionary trajectories that produced it some 500 million years ago. While we cannot make direct observations on extinct species, we can examine the many analogous global broadcasts, neural and otherwise, rapid and otherwise, available to empirical study within and across many individual organisms, species, taxa, and their interacting communities.

This chapter argues that the ubiquity of cognitive process at all levels of scale and organization within the living state leaves open to evolutionary exaptation the inevitable occurrence of both crosstalk and noise. A principal outcome of such exaptation will be the repeated evolution of punctuated global broadcast mechanisms that entrain sets of "unconscious" cognitive modules into shifting, tunable cooperative arrays tasked with meeting the changing patterns of threat and opportunity that challenge all organisms. When such entrainment involves neural systems acting on a timescale of a few hundred milliseconds, the phenomenon is characterized as consciousness. When entrainment involves many individuals or cultural artifacts, the outcomes are social or institutional processes. All such phenomena will, of course, also interact by crosstalk. Following the injunction of Huys et al. (2016) this is a mechanism "…linking multiple interacting levels, from molecules to cells, circuits, cognition, behavior, and the physical and social environment."

References

Adams, F. 2003. The informational turn in philosophy. *Minds and Machines* 13: 471–501.

Ash, R. 1990. *Information Theory*. New York: Dover.

Atiyah, M., and I. Singer. 1963. The index of elliptical operators on compact manifolds. *Bulletin of the American Mathematical Society* 69: 322–433.

Atlan, H., and I. Cohen. 1998. Immune information, self-organization, and meaning. *International Immunology* 10: 711–717.

Baars, B. 1988. *A Cognitive Theory of Consciousness*. New York: Cambridge University Press.

Baars, B. 2005. Global workspace theory of consciousness: Toward a cognitive neuroscience of human experience. *Progress in Brain Research* 150: 45–53.

Baars, B., and S. Franklin. 2003. How conscious experience and working memory interact. *Trends in Cognitive Science* 7: 166–172.

Backdahl, L., A. Bushell, and S. Beck. 2009. Inflammatory signalling as mediator of epigenetic modulation in tissue-specific chronic inflammation. *The International Journal of Biochemistry and Cell Biology* 41: 176–184.

Beck, C., and F. Schlogl. 1995. *Thermodynamics of Chaotic Systems*. New York: Cambridge University Press.

Bennett, C. 1988. Logical depth and physical complexity. In *The Universal Turing Machine: A Half-Century Survey*, ed. R. Herkin, 227–257. New York: Oxford University Press.

Bennett, M., and P. Hacker. 2003. *Philosophical Foundations of Neuroscience*. London: Blackwell Publishing.

Boccaletti, S., V. Latora, Y. Moreno, M. Chavez, and D. Hwang. 2006. Complex networks: Structure and dynamics. *Physics Reports* 424: 175–208.

Bonner, J. 1980. *The Evolution of Culture in Animals*. Princeton, NJ: Princeton University Press.

Byers, N. 1999. E. Noether's discovery of the deep connection between symmetries and conservation laws. In *Israel Mathematical Conference Proceedings*, vol. 12. ArXiv physics/9807044.

Champagnet, N., R. Ferriere, and S. Meleard. 2006. Unifying evolutionary dynamics: From individual stochastic process to macroscopic models. *Theoretical Population Biology* 69: 297–321.

Cohen, I. 1992. The cognitive principle challenges clonal selection. *Immunology Today* 13: 441–444.

Cohen, I. 2000. *Tending Adam's Garden: Evolving the Cognitive Immune Self.* New York: Academic.

Cohen, I. 2006. Immune system computation and the immunological homunculus. In *MoDels 2006*, ed. O. Nierstrasz, J. Whittle, D. Harel, and G. Reggio. LCNS, vol. 4199, 499–512. Heidelberg: Springer.

Cohen, I., and D. Harel. 2007. Explaining a complex living system: Dynamics, multiscaling, and emergence. *Journal of the Royal Society: Interface* 4: 175–182.

Cover, T., and J. Thomas. 2006. *Elements of Information Theory*, 2nd ed. New York: Wiley.

de Groot, S., and P. Mazur. 1984. *Non-equilibrium Thermodynamics*. New York: Dover.

Dembo, A., and O. Zeitouni. 1998. *Large Deviations: Techniques and Applications*, 2nd ed. New York: Springer.

Dretske, F. 1981. *Knowledge and the Flow of Information*. Cambridge, MA: MIT Press.

Dretske, F. 1988. *Explaining Behavior*. Cambridge, MA: MIT Press.

Dretske, F. 1992. What isn't wrong with folk psychology. *Metaphilosophy* 29: 1–13.

Dretske, F. 1993. Mental events as structuring causes of behavior. In *Mental Causation*, ed. A. Mele and J. Heil, 121–136. New York: Oxford University Press.

Dretske, F. 1994. The explanatory role of information. *Philosophical Transactions of the Royal Society A* 349: 59–70.

Durham, W. 1991. *Coevolution: Genes, Culture, and Human Diversity*. Palo Alto, CA: Stanford University Press.

El Gamal, A., and Y. Kim, 2010. Lecture notes on network information theory. ArXiv:1001.3404v4.
Ellis, R. 1985. *Entropy, Large Deviations, and Statistical Mechanics*. New York: Springer.
English, T. 1996. Evaluation of evolutionary and genetic optimizers: No free lunch. In *Evolutionary Programming V: Proceedings of the Fifth Annual Conference on Evolutionary Programming*, ed. L. Fogel, P. Angeline, and T. Back, 163–169. Cambridge, MA: MIT Press.
Erdos, P., and A. Renyi. 1960. On the evolution of random graphs. *Publication of the Mathematical Institute of the Hungarian Academy of Sciences* 5: 17–61.
Feynman, R. 2000. *Lectures on Computation*. New York: Westview Press.
Foley, D., J. Craid, R. Morley, C. Olsson, T. Dwyer, K. Smith, and R. Saffery. 2009. Prospects for epigenetic epidemiology. *American Journal of Epidemiology* 169: 389–400.
Forlenza, M., and A. Baum. 2000. Psychosocial influences on cancer progression: Alternative cellular and molecular mechanisms. *Current Opinion in Psychiatry* 13: 639–645.
Glazebrook, J.F., and R. Wallace. 2009a. Rate distortion manifolds as model spaces for cognitive information. *Informatica* 33: 309–346.
Glazebrook, J.F., and R. Wallace. 2009b. Small worlds and red queens in the global workspace: An information-theoretic approach. *Cognitive Systems Research* 10: 333–365.
Gould, S. 2002. *The Structure of Evolutionary Theory*. Cambridge, MA: Harvard University Press.
Gould, S., and R. Lewontin. 1979. The spandrels of San Marco and the Panglossian paradigm: A critique of the adaptationist programme. *Proceedings of the Royal Society of London, B* 205: 581–598.
Haggstrom, M. 2012. Wikipedia entry on 'wound healing'.
Hazewinkel, M. 2002. *Encyclopedia of Mathematics, 'Index Formulas'*. New York: Springer.
Hollan, J., J. Hutchins, and D. Kirsch. 2000. Distributed cognition: Toward a new foundation for human-computer interaction research. *ACM Transactions on Computer-Human Interaction* 7: 174–196.
Huys, Q., T. Maia, and M. Frank. 2016. Computational psychiatry as a bridge from neuroscience to clinical applications. *Nature Neuroscience* 19: 404–413.
Jablonka, E. 2004. Epigenetic epidemiology. *International Journal of Epidemiology* 33: 929–935.
Jablonka, E., and M. Lamb. 1995. *Epigenetic Inheritance and Evolution: The Lamarckian Dimension*. Oxford: Oxford University Press.
Jablonka, E., and M. Lamb. 1998. Epigenetic inheritance in evolution. *Journal of Evolutionary Biology* 11: 159–183.
Jaenisch, R., and A. Bird. 2003. Epigenetic regulation of gene expression: How the genome integrates intrinsic and environmental signals. *Nature Genetics Supplement* 33: 245–254.
Khinchin, A. 1957. *The Mathematical Foundations of Information Theory*. New York: Dover.
Kitano, H. 2004. Biological robustness. *Nature Reviews Genetics* 5: 826–837.
Landau, L., and E. Lifshitz. 2007. *Statistical Physics*, 3rd ed. Part I. Elsevier, New York.
Maturana, H., and F. Varela. 1980. *Autopoiesis and Cognition*. Dordrecht: Reidel Publishing Company.
Maturana, H., and F. Varela. 1992. *The Tree of Knowledge*. Boston, MA: Shambhala Publications.
Mindwood, K., L. Valenick Williams, and J. Schwarzbauer. 2004. Tissue repair and the dynamics of the extracellular matrix. *International Journal of Biochemistry and Cell Biology* 36: 1031–1037.
Nocedal, J., and S. Wright. 1999. *Numerical Optimization*. New York: Springer.
Nunney, L. 1999. Lineage selection and the evolution of multistage carcinogenesis. *Proceedings of the Royal Society B* 266: 493–498.
O'Nuallain, S. 2008. Code and context in gene expression, cognition, and consciousness, Chap. 15. In *The Codes of Life: The Rules of Macroevolution*, ed. M. Barbiere, 347–356. New York: Springer.

Pettini, M. 2007. *Geometry and Topology in Hamiltonian Dynamics and Statistical Mechanics*. New York: Springer.

Protter, P. 1990. *Stochastic Integration and Differential Equations*. New York: Springer.

Richerson, P., and R. Boyd. 2006. *Not by Genes Alone: How Culture Transformed Human Evolution*. Chicago, IL: Chicago University Press.

Rockafellar, R. 1970. *Convex Analysis*. Princeton, NJ: Princeton University Press.

Sergent, C., and S. Dehaene. 2004. Is consciousness a gradual phenomenon? Evidence for an all-or-none bifurcation during the attentional blink. *Psychological Science* 15: 720–725.

Scherrer, K., and J. Jost. 2007a. The gene and the genon concept: A functional and information-theoretic analysis. *Molecular Systems Biology* 3: 87–95.

Scherrer, K., and J. Jost. 2007b. Gene and genon concept: Coding versus regulation. *Theory in Bioscience* 126: 65–113.

Shannon, C. 1959. Coding theorems for a discrete source with a fidelity criterion. *Institute of Radio Engineers International Convention Record* 7: 142–163.

Spenser, J. 2010. The giant component: The golden anniversary. *Notices of the American Mathematical Society* 57: 720–724.

Tegmark, M. 2000. Importance of quantum decoherence in brain processes. *Physical Review E* 61: 4194–4206.

Turner, B. 2000. Histone acetylation and an epigenetic code. *Bioessays* 22: 836–845.

Wallace, R. 2000. Language and coherent neural amplification in hierarchical systems: Renormalization and the dual information source of a generalized spatiotemporal stochastic resonance. *International Journal of Bifurcation and Chaos* 10: 493–502.

Wallace, R. 2005. *Consciousness: A Mathematical Treatment of the Global Neuronal Workspace Model*. New York: Springer.

Wallace, R. 2006. Pitfalls in biological computing: Canonical and idiosyncratic dysfunctions of conscious machines. *Mind and Matter* 4: 91–113.

Wallace, R. 2007. Culture and inattentional blindness: A global workspace perspective. *Journal of Theoretical Biology* 245: 378–390.

Wallace, R. 2008. Toward formal models of biologically inspired, highly parallel machine cognition. *International Journal of Parallel, Emergent and Distributed Systems* 23: 367–408.

Wallace, R. 2009. Programming coevolutionary machines: The emerging conundrum. *International Journal of Parallel, Emergent and Distributed Systems* 24: 443–453.

Wallace, R. 2010. Expanding the modern synthesis. *Comptes Rendus Biologies* 333: 701–709.

Wallace, R. 2017. *Information Theory Models of Instabilities in Critical Systems*. World Scientific Series in Information Studies, vol. 7. Singapore: World Scientific.

Wallace, R., and M. Fullilove. 2008. *Collective Consciousness and Its Discontents*. New York: Springer.

Wallace, R., and D. Wallace. 2008. Punctuated equilibrium in statistical models of generalized coevolutionary resilience: How sudden ecosystem transitions can entrain both phenotype expression and Darwinian selection. *Transactions on Computational Systems Biology IX*, LNBI. 5121: 23–85.

Wallace, R., and D. Wallace. 2009. Code, context, and epigenetic catalysis in gene expression. *Transactions on Computational Systems Biology XI*, LNBI. 5750: 283–334.

Wallace, R., and D. Wallace. 2010. *Gene Expression and Its Discontents: The Social Production of Chronic Disease*. New York: Springer.

Wallace, R., and D. Wallace. 2013. *A Mathematical Approach to Multilevel, Multiscale Health Interventions: Pharmaceutical Industry Decline and Policy Response*. London: Imperial College Press.

Wallace, R., D. Wallace, and R.G. Wallace. 2003. Toward cultural oncology: The evolutionary information dynamics of cancer. *Open Systems and Information Dynamics* 10: 159–181.

Wilson, K. 1971. Renormalization group and critical phenomena. I renormalization group and the Kadanoff scaling picture. *Physics Reviews B* 4: 3174–3183.

Wolpert, D., and W. MacReady. 1995. No free lunch theorems for search, Santa Fe Institute, SFI-TR-02-010.

Wolpert, D., and W. MacReady. 1997. No free lunch theorems for optimization. *IEEE Transactions on Evolutionary Computation* 1: 67–82.

Zhu, R., A. Rebirio, D. Salahub, and S. Kaufmann. 2007. Studying genetic regulatory networks at the molecular level: Delayed reaction stochastic models. *Journal of Theoretical Biology* 246: 725–745.

Chapter 2
Cultural Epigenetics: On the Missing Heritability of Complex Diseases

Written in collaboration with Dr. D.N. Wallace.

Summary We extend a cognitive paradigm for gene expression based on the asymptotic limit theorems of information theory to the epigenetic epidemiology of complex developmental disorders in humans. In particular, we recognize the fundamental role culture plays in human biology, another heritage mechanism parallel to, and interacting with, the more familiar genetic and epigenetic systems. We do this via a model through which culture acts as another tunable epigenetic catalyst that both directs developmental trajectories, and becomes convoluted with individual ontology, via a mutually interacting crosstalk mediated by a social interaction that is itself culturally driven. We call for the incorporation of embedding culture as an essential component of the epigenetic regulation of human development and its dysfunctions. The cultural and epigenetic systems of heritage may well provide the "missing" heritability of complex diseases now under so much intense discussion.

2.1 Introduction

Mental Disorders and Culture

We begin with a discussion of human mental disorders, that, while increasingly recognized as quintessentially developmental in nature, are still not well understood. The classic scientific task will then be to infer the general from the particular, extending focus on the central role of culture in human mental dysfunction to a vastly larger spectrum of developmental pathologies. This is a task that requires cutting-edge mathematical methods for its formal expression.

The understanding of mental disorders is in considerable disarray. Official classifications of mental illness such as the *Diagnostic and Statistical Manual of Mental Disorders - fourth edition*, (DSM-IV 1994, and now its replacement as "DSM-V"), the standard descriptive nosology in the USA, have even been characterized as "prescientific" by P. Gilbert (2001) and others. Johnson-Laird et al. (2006) claim that current knowledge about psychological illnesses is comparable to the medical understanding of epidemics in the early nineteenth century. Physicians realized then that cholera, for example, was a specific disease, which killed about a third of the people whom it infected. What they disagreed about was the cause, the pathology,

R. Wallace, *Computational Psychiatry*, DOI 10.1007/978-3-319-53910-2_2

and the communication of the disease. Similarly, according to Johnson-Laird et al., most medical professionals these days realize that psychological illnesses occur (cf. DSMIV), but they disagree about their cause and pathology. Notwithstanding DSMIV, Johnson-Laird et al. doubt whether any satisfactory a priori definition of psychological illness can exist because it is a matter for theory to elucidate.

Atmanspacher (2006) argues that formal theory of high-level cognitive process is itself at a point similar to that of physics 400 years ago, in that the basic entities, and the relations between them, have yet to be delineated.

More generally, simple arguments from genetic determinism regarding mental disorders fail, in part because of a draconian population bottleneck that, early in our species' history, resulted in an overall genetic diversity less than that observed within and between contemporary chimpanzee subgroups.

Manolio et al. (2009) describe this conundrum more generally in terms of "finding the missing heritability of complex diseases." They observe, for example, that at least 40 loci have been associated with human height, a classic complex trait with an estimated heritability of about 80%, yet they explain only about 5% of phenotype variance despite studies of tens of thousands of people. This result, they find, is typical across a broad range of supposedly heritable diseases, and call for extending beyond current genome-wide association approaches to illuminate the genetics of complex diseases and enhance its potential to enable effective disease prevention or treatment.

Arguments from psychosocial stress fare better (e.g., Brown et al. 1973; Dohrenwend and Dohrenwend 1974; Eaton 1978), particularly for depression (e.g., Risch et al. 2009), but are affected by the apparently complex and contingent developmental paths determining the onset of schizophrenia, dementias, psychoses, and so forth, some of which may be triggered in utero by exposure to infection, low birth weight, or other functional teratogens.

P. Gilbert (2001) suggests an extended evolutionary perspective, in which evolved mechanisms like the "flight-or-fight" response are inappropriately excited or suppressed, resulting in such conditions as anxiety or post-traumatic stress disorders. Nesse (2000) suggests that depression may represent the dysfunction of an evolutionary adaptation which downregulates foraging activity in the face of unattainable goals.

Kleinman and Good, however, (1985, p. 492) outline something of the cross cultural subtleties affecting the study of depression that seem to argue against any simple evolutionary or genetic interpretation. They state that, when culture is treated as a constant, as is common when studies are conducted in our own society, it is relatively easy to view depression as a biological disorder, triggered by social stressors in the presence of ineffective support, and reflected in a set of symptoms or complaints that map back onto the biological substrate of the disorder. However, they continue, when culture is treated as a significant variable, for example, when the researcher seriously confronts the world of meaning and experience of members of non-Western societies, many of our assumptions about the nature of emotions and illness are cast in sharp relief.

Dramatic differences are found across cultures in the social organization, personal experience, and consequences of such emotions as sadness, grief, and anger, of behaviors such as withdrawal or aggression, and of psychological characteristics such as passivity and helplessness or the resort to altered states of consciousness. They are organized differently as psychological realities, communicated in a wide range of idioms, related to quite varied local contexts of power relations, and are interpreted, evaluated, and responded to as fundamentally different meaningful realities. Depressive illness and dysphoria are thus not only interpreted differently in non-Western societies and across cultures; they are *constituted* as fundamentally different forms of social reality.

Since the publication of that landmark study, a number of comprehensive overviews have been published that support its conclusions, for example, Bebbington (1993), Jenkins et al. (1990), Journal of Clinical Psychiatry (Supplement 13), and Manson (1995). As Marsella (2003) writes, it is now clear that cultural variations exist in the areas of meaning, perceived causes, onset patterns, epidemiology, symptom expression, course, and outcome, variations having important implications for understanding clinical activities including conceptualization, assessment, and therapy.

Kleinman and Cohen (1997) argue forcefully that several myths have become central to psychiatry. The first is that the forms of mental illness everywhere display similar degrees of prevalence. The second is an excessive adherence to a principle known as the pathogenic/pathoplastic dichotomy, which holds that biology is responsible for the underlying structure of a malaise, whereas cultural beliefs shape the specific ways in which a person experiences it. The third myth maintains that various unusual culture-specific disorders whose biological bases are uncertain occur only in exotic places outside the West. In an effort to base psychiatry in "hard" science and thus raise its status to that of other medical disciplines, psychiatrists have narrowly focused on the biological underpinnings of mental disorders while discounting the importance of such "soft" variables as culture and socioeconomic status.

Heine (2001) describes an explicit cultural psychology that views the person as containing a set of biological potentials interacting with particular situational contexts that constrain and afford the expression of various constellations of traits and patterns of behavior. He says that, unlike much of personality psychology, cultural psychology focuses on the constraints and affordances inherent to the cultural environment that give shape to those biological potentials. Human nature, from this perspective, is seen as emerging from participation in cultural worlds, and of adapting oneself to the imperatives of cultural directives, meaning that our nature is ultimately that of a cultural being.

Heine describes how cultural psychology does not view culture as a superficial wrapping of the self, or as a framework within which selves interact, but as something that is intrinsic to the self, so that without culture there is no self, only a biological entity deprived of its potential. Individual selves, from Heine's perspective, are inextricably grounded in a configuration of consensual understandings and behavioral customs particular to a given cultural and historical context, so that understanding the self requires an understanding of the culture that sustains it. Heine argues, then, that the process of becoming a self is contingent on individuals interacting with, and seizing meanings from, the cultural environment.

Heine warns that the extreme nature of American individualism means that a psychology based on the late twentieth century American research not only stands the risk of developing models that are particular to that culture, but also of developing an understanding of the self that is peculiar in the context of the world's cultures.

Indeed, as Norenzayan and Heine (2005) point out, for the better part of a century, a considerable controversy has raged within anthropology regarding the degree to which psychological and other human universals do, in fact, actually exist independent of the particularities of culture.

Many other observers have made similar points over the years (e.g., Arnett 2008; Henrich et al. 2010; Markus and Kitayama 1991; Matsuda and Nisbett 2006; Nisbett et al. 2001; Wallace 2007).

As Durham (1991) and Richerson and Boyd (2004) explore at some length, humans are endowed with two distinct but interacting heritage systems: genes and culture. Durham (1991), for example, writes that genes and culture constitute two distinct but interacting systems of information inheritance within human populations and information of both kinds has influence, actual or potential, over behaviors, which creates a real and unambiguous symmetry between genes and phenotypes on the one hand, and culture and phenotypes on the other. Genes and culture, in his view, are best represented as two parallel lines or tracks of hereditary influence on phenotypes.

Both genes and culture can be envisioned as generalized languages in that they have recognizable "grammar" and "syntax," in the sense of Wallace (2005) and Wallace and Wallace (2008, 2009).

More recent work has identified epigenetic heritage mechanisms involving such processes as environmentally induced gene methylation, that can have strong influence across several generations (e.g., Jablonka and Lamb 1995, 1998; Jablonka 2004; Wallace and Wallace 2009), and are the subject of intense current research, a matter to which we will return below.

There are, it seems, two powerful heritage mechanisms in addition to the genetic where one may perhaps find the "missing heritability of complex diseases" that Manolio et al. seek.

Here we will expand recent explorations of a cognitive paradigm for gene expression (Wallace and Wallace 2008, 2009) that incorporates the effects of surrounding epigenetic regulatory machinery as a kind of catalyst to include the effects of the embedding information source of human culture on human ontology. The essential feature is that a cognitive process, including gene expression, can instantiate a dual information source that can interact with the generalized language of culture in which, for example, social interplay and the interpretation of socioeconomic and environmental stressors involve complicated matters of symbolism and its grammar and syntax. These information sources interact by a crosstalk that, over the life course, determines human ontology and its manifold dysfunctions.

2.2 A Cognitive Paradigm for Gene Expression

Recapitulating something of the previous chapter, a cognitive paradigm for gene expression is under active study, a model in which contextual factors determine the behavior of what Cohen characterizes as a "reactive system," not at all a deterministic—or even simple stochastic—mechanical process (e.g., Cohen 2006; Cohen and Harel 2007; Wallace and Wallace 2008, 2009). The various formal approaches are, however, all in the spirit of Maturana and Varela (1980, 1992) who understood the central role that cognitive process must play across a vast array of biological phenomena.

O'Nuallain (2008) puts gene expression directly in the realm of complex linguistic behavior, for which context imposes meaning. He claims that the analogy between gene expression and language production is useful both as a fruitful research paradigm and also, given the relative lack of success of natural language processing (nlp) by computer, as a cautionary tale for molecular biology. A relatively simple model of cognitive process as an information source permits use of Dretske's (1994) insight that any cognitive phenomenon must be constrained by the limit theorems of information theory, in the same sense that sums of stochastic variables are constrained by the Central Limit Theorem. This perspective permits a new formal approach to gene expression and its dysfunctions, in particular suggesting new and powerful statistical tools for data analysis that could contribute to exploring both ontology and its pathologies. Wallace and Wallace (1998, 2009) apply the perspective, respectively, to infectious and chronic disease. Here we extend the mathematical foundations of that work to examine the topological structures of development and developmental disorder, in the context of an embedding information source representing the compelling varieties of human culture.

This approach is consistent with the broad context of epigenetics and epigenetic epidemiology. Jablonka and Lamb (1995, 1998), for example, argue that information can be transmitted from one generation to the next in ways other than through the base sequence of DNA. It can be transmitted through cultural and behavioral means in higher animals, and by epigenetic means in cell lineages. All of these transmission systems allow the inheritance of environmentally induced variation. Such Epigenetic Inheritance Systems are the memory systems that enable somatic cells of different phenotypes but identical genotypes to transmit their phenotypes to their descendants, even when the stimuli that originally induced these phenotypes are no longer present.

This epigenetic perspective has received much empirical confirmation (e.g., Backdahl et al. 2009; Turner 2000; Jaenisch and Bird 2003; Jablonka 2004).

Foley et al. (2009) argue that epimutation is estimated to be 100 times more frequent than genetic mutation and may occur randomly or in response to the environment. Periods of rapid cell division and epigenetic remodeling are likely to be most sensitive to stochastic or environmentally mediated epimutation. Disruption of epigenetic profile is a feature of most cancers and is speculated to play a role in the etiology of other complex diseases including asthma allergy, obesity, type 2 diabetes, coronary heart disease, autism spectrum disorders and bipolar disorders, and schizophrenia.

Similarly, Scherrer and Jost (2007a,b) explicitly invoke information theory in their extension of the definition of the gene to include the local epigenetic machinery, a construct they term the "genon." Their point is that coding information is not simply contained in the coded sequence, but is, in their terms, *provided by* the genon that accompanies it on the expression pathway and controls in which peptide it will end up. In their view the information that counts is not about the identity of a nucleotide or an amino acid derived from it, but about the relative frequency of the transcription and generation of a particular type of coding sequence that then contributes to the determination of the types and numbers of functional products derived from the DNA coding region under consideration.

Again, the proper formal tools for understanding phenomena that "provide" information—that are information sources—are the Rate Distortion Theorem and its zero error limit, the Shannon–McMillan Theorem.

2.3 Models of Development

The currently popular spinglass model of development (e.g., Ciliberti et al. 2007a,b) assumes that N transcriptional regulators are represented by their expression patterns

$$\mathbf{S}(t) = [S_1(t), \ldots, S_N(t)] \tag{2.1}$$

at some time t during a developmental or cell-biological process and in one cell or domain of an embryo. The transcriptional regulators influence each other's expression through cross-regulatory and autoregulatory interactions described by a matrix $w = (w_{ij})$. For nonzero elements, if $w_{ij} > 0$ the interaction is activating, if $w_{ij} < 0$ it is repressing. w represents, in this model, the regulatory genotype of the system, while the expression state $\mathbf{S}(t)$ is the phenotype. These regulatory interactions change the expression of the network $\mathbf{S}(t)$ as time progresses according to a difference equation

$$S_i(t + \Delta t) = \sigma \left[\sum_{j=1}^{N} w_{ij} S_j(t) \right] \tag{2.2}$$

where Δt is a constant and σ a sigmoidal function whose value lies in the interval $(-1, 1)$. In the spinglass limit σ is the sign function, taking only the values ± 1.

The regulatory networks of interest here are those whose expression state begins from a prespecified initial state $\mathbf{S}(0)$ at time $t = 0$ and converges to a prespecified stable equilibrium state $\mathbf{S}_\infty$. Such networks are termed *viable* and must necessarily be a very small fraction of the total possible number of networks, since most do not begin and end on the specified states. This "simple" observation is not at all simple in our reformulation, although other results become far more accessible, as we can then invoke the asymptotic limit theorems of information theory.

The spinglass approach to development is formally similar to spinglass neural network models of learning by selection, e.g., as proposed by Toulouse et al. (1986) nearly a generation ago. Much subsequent work, summarized by Dehaene and Naccache (2001), suggests that such models are simply not sufficient to the task of understanding high-level cognitive function, and these have been largely supplanted by complicated "global workspace" concepts whose mathematical characterization is highly nontrivial (Atmanspacher 2006).

It is possible to shift the perspective on development by invoking a cognitive paradigm for gene expression, following the example of the Atlan/Cohen model of immune cognition.

Atlan and Cohen (1998), in the context of a study of the immune system, argue that the essence of cognition is the comparison of a perceived signal with an internal (learned or inherited) picture of the world, and then, upon that comparison, the choice of a single response from a larger repertoire of possible responses.

Such choice inherently involves information and information transmission since it always generates a reduction in uncertainty, as explained by Ash (1990, p. 21).

More formally, a pattern of incoming input—like the $\mathbf{S}(t)$ above—is mixed in a systematic algorithmic manner with a pattern of internal ongoing activity—like the (w_{ij}) above—to create a path of combined signals $x = (a_0, a_1, \ldots, a_n, \ldots)$—analogous to the sequence of $\mathbf{S}(t + \Delta t)$ above, with, say, $n = t/\Delta t$. Each a_k thus represents some functional composition of internal and external signals.

For a cognitive process, this path is supposed to be fed into a "highly nonlinear decision oscillator," h, a sudden threshold machine whose canonical model could well be taken as the famous integrate-and-fire neuron (e.g., Hoppensteadt and Izhikevich 1997, Proposition 8.12). $h(x)$, otherwise seen as a "black box," thus generates an output that is an element of one of two disjoint sets B_0 and B_1 of possible system responses. Let us define the sets B_k as

$$B_0 \equiv \{b_0, \ldots, b_k\}$$

$$B_1 \equiv \{b_{k+1}, \ldots, b_m\} \tag{2.3}$$

Assume a graded response, supposing that if $h(x) \in B_0$, the pattern is not recognized, and if $h(x) \in B_1$, the pattern has been recognized, and some action $b_j, k + 1 \leq j \leq m$ takes place.

Rather than focusing on the properties of h, *we shift the perspective*: The principal objects of formal interest become *paths* x triggering pattern recognition-and-response. That is, given a fixed initial state a_0, examine all possible subsequent paths x beginning with a_0 and leading to the event $h(x) \in B_1$. Thus $h(a_0, \ldots, a_j) \in B_0$ for all $0 < j < m$, but $h(a_0, \ldots, a_m) \in B_1$.

Several points are central to the shift in perspective we are making:

1. It is very important to understand that the fundamental core of the argument does not regard the exact internal details of the inferred (but perhaps not easily observable) function $h(x)$, but rather in the "grammar" and "syntax" of the strings $x = a_0, a_1, \ldots$ leading to action of that function, and which are more likely to be observable.
2. For each positive integer n, let $N(n)$ be the number of high probability grammatical and syntactical paths of length n which begin with some particular a_0 and lead to the condition $h(x) \in B_1$. Call such paths "meaningful," assuming, not unreasonably, that $N(n)$ will be considerably less than the number of all possible paths of length n leading from a_0 to the condition $h(x) \in B_1$.
3. While the combining algorithm, the form of the "nonlinear oscillator" h, and the details of grammar and syntax are all unspecified in this model, *the critical assumption* that permits inference of the necessary conditions constrained by the asymptotic limit theorems of information theory is that the finite limit

$$H \equiv \lim_{n \to \infty} \frac{\log[N(n)]}{n} \tag{2.4}$$

both exists and is independent of the path x.

Define such a pattern recognition-and-response cognitive process as *ergodic*. Not all cognitive processes are likely to be ergodic in this sense, implying that H, if it indeed exists at all, is path dependent, although extension to nearly ergodic processes seems possible (Wallace and Fullilove 2008).

Invoking the spirit of the Shannon–McMillan Theorem, as choice involves an inherent reduction in uncertainty, it is then possible to define an adiabatically, piecewise stationary, ergodic (APSE) information source $\mathbf{X}$ associated with stochastic variates X_j having joint and conditional probabilities $P(a_0, \ldots, a_n)$ and $P(a_n | a_0, \ldots, a_{n-1})$ such that appropriate conditional and joint Shannon uncertainties satisfy the classic relations

$$\begin{aligned} H[\mathbf{X}] &= \lim_{n \to \infty} \frac{\log[N(n)]}{n} \\ &= \lim_{n \to \infty} H(X_n | X_0, \ldots, X_{n-1}) \\ &= \lim_{n \to \infty} \frac{H(X_0, \ldots, X_n)}{n+1} \end{aligned} \tag{2.5}$$

This information source is defined as *dual* to the underlying ergodic cognitive process.

Adiabatic means that the source has been parameterized according to some scheme, and that, over a certain range, along a particular piece, as the parameters vary, the source remains as close to stationary and ergodic as needed for information theory's central theorems to apply. *Stationary* means that the system's probabilities do not change in time, and *ergodic*, roughly, that the cross-sectional means approximate long-time averages. Between pieces it is necessary to invoke various kinds of phase transition formalisms, as described more fully in Wallace (2005) or Wallace and Wallace (2008).

In the developmental vernacular of Ciliberti et al., we now examine paths in phenotype space that begin at some $\mathbf{S}_0$ and converge $n = t/\Delta t \rightarrow \infty$ to some other $\mathbf{S}_\infty$. Suppose the system is conceived at $\mathbf{S}_0$, and h represents (for example) reproduction when phenotype $\mathbf{S}_\infty$ is reached. Thus $h(x)$ can have two values, i.e., B_0 not able to reproduce, and B_1, mature enough to reproduce. Then $x = (\mathbf{S}_0, \mathbf{S}_{\Delta t}, \ldots, \mathbf{S}_{n\Delta t}, \ldots)$ until $h(x) = B_1$.

Structure is now subsumed *within the sequential grammar and syntax of the dual information source* rather than within the cross-sectional internals of (w_{ij})-space, a simplifying shift in perspective.

This transformation carries considerable computational burdens, as well as, and perhaps in consequence of, providing deep mathematical insight.

First, the fact that viable networks comprise a tiny fraction of all those possible emerges easily from the spinglass formulation simply because of the "mechanical" limit that the number of paths from $\mathbf{S}_0$ to $\mathbf{S}_\infty$ will always be far smaller than the total number of possible paths, most of which simply do not end on the target configuration.

From the information source perspective, which inherently subsumes a far larger set of dynamical structures than possible in a spinglass model—not simply those of symbolic dynamics—the result is what Khinchin (1957) characterizes as the "E-property" of a stationary, ergodic information source. This property allows, in the limit of infinitely long output, the classification of output strings into two sets:

1. a very large collection of gibberish which does not conform to underlying (sequential) rules of grammar and syntax, in a large sense, and which has near-zero probability, and
2. a relatively small "meaningful" set, in conformity with underlying structural rules, having very high probability.

The essential content of the Shannon–McMillan Theorem is that, if $N(n)$ is the number of meaningful strings of length n, then the uncertainty of an information source X can be defined as

$$H[X] = \lim_{n \to \infty} \log[N(n)]/n$$

that can be expressed in terms of joint and conditional probabilities. Proving these results for general stationary, ergodic information sources require considerable mathematical machinery (e.g., Khinchin 1957; Cover and Thomas 2006; Dembo and Zeitouni 1998).

Second, according to Ash (1990), information source uncertainty has an important heuristic interpretation in that we may regard a portion of text in a particular language as being produced by an information source. A large uncertainty means, by the Shannon–McMillan Theorem, a large number of "meaningful" sequences. Thus given two languages with uncertainties H_1 and H_2, respectively, if $H_1 > H_2$, then in the absence of noise it is easier to communicate in the first language; more can be said in the same amount of time. On the other hand, it will be easier to reconstruct a scrambled portion of text in the second language, since fewer of the possible sequences of length n are meaningful.

Third, information source uncertainty is homologous with free energy density in a physical system, a matter having implications across a broad class of dynamical behaviors.

The free energy density of a physical system having volume V and partition function $Z(K)$ derived from the system's Hamiltonian—the energy function—at inverse temperature K is (e.g., Landau and Lifshitz 2007)

$$
\begin{aligned}
F[K] &= \lim_{V\to\infty} -\frac{1}{K}\frac{\log[Z(K,V)]}{V} \\
&= \lim_{V\to\infty} \frac{\log[\hat{Z}(K,V)]}{V}
\end{aligned}
\tag{2.6}
$$

where $\hat{Z} = Z^{-1/K}$.

The partition function for a physical system is the normalizing sum in an equation having the form

$$
P[E_i] = \frac{\exp[-E_i/kT]}{\sum_j \exp[-E_j/kT]} \tag{2.7}
$$

where E_i is the energy of state i, k a constant, and T the system temperature.

Feynman (2000), following the classic approach by Bennett (1988), who examined idealized machines using information to do work, concludes that *the information contained in a message is most simply measured by the free energy needed to erase it.*

Thus, according to this argument, source uncertainty is homologous to free energy density as defined above, i.e., from the similarity with the relation $H = \lim_{n\to\infty} \log[N(n)]/n$.

Ash's perspective then has an important corollary: If, for a biological system, $H_1 > H_2$, source 1 will require more metabolic free energy than source 2.

2.4 Tunable Epigenetic Catalysis

Incorporating the influence of embedding contexts—generalized epigenetic effects—is most elegantly done by invoking the Joint Asymptotic Equipartition Theorem (JAEPT) (Cover and Thomas 2006). For example, given an embedding epigenetic information source, say Y, that affects development, then the dual cognitive source uncertainty $H[X]$ is replaced by a joint uncertainty $H(X,Y)$.

The objects of interest then become the jointly typical dual sequences $z^n = (x^n, y^n)$, where x is associated with cognitive gene expression and y with the embedding epigenetic regulatory context. Restricting consideration of x and y to those sequences that are in fact jointly typical allows use of the information transmitted from Y to X as the splitting criterion.

From the information theory "chain rule" (Cover and Thomas 2006),

$$H(X, Y) = H(X) + H(Y|X) \leq H(X) + H(Y)$$

Equality occurs only for stochastically independent processes.

Interpreting the homology between information and free energy rather broadly, the embedding context, in effect *lowers an analog to the relative activation energy of a particular developmental channel*, at the expense of raising the total free energy needed, since the system must now support two information sources instead of one, i.e., that regulated, and that providing the regulation.

Thus the effect of epigenetic regulation is to change the probability of developmental pathways, while requiring more total energy for development. Hence the epigenetic information source Y acts as a *tunable catalyst*, a kind of second order cognitive enzyme, to enable and direct developmental pathways. This result permits hierarchical models similar to those of higher order cognitive neural function that incorporate contexts in a natural way (e.g., Wallace and Wallace 2008; Wallace and Fullilove 2008). The cost of this ability to channel is the metabolic necessity of supporting two information sources, X and Y, rather than just X itself.

This elaboration allows a spectrum of possible "final" phenotypes, what S. Gilbert (2001) calls developmental or phenotype plasticity. Thus gene expression is seen as, in part, responding to environmental or other, internal, developmental signals.

Including the effects of embedding culture in human ontology is, according to this formalism, straightforward: Consider culture as another embedding information source, Z, having source uncertainty $H(Z)$. Then the information chain rule becomes

$$H(X, Y, Z) \leq H(X) + H(Y) + H(Z) \tag{2.8}$$

and the numbers of "typical" sequences of length n are then approximately

$$\exp[nH(X, Y, Z)] \leq \exp[n(H(X) + H(Y) + H(Z))] \tag{2.9}$$

where, again, equality occurs only under stochastic independence.

A cultural regulatory apparatus, however, has very considerable free energy requirements, to grossly understate the matter.

In this model, following explicitly the direction indicated by Boyd, Kleinman, and their colleagues, culture is seen as an essential component of the catalytic epigenetic machinery that regulates human ontology, including development of the human mind. This is not to say that the development in other animals, particularly those that are highly social, does not undergo analogous regulation by larger-scale structures of interaction. For human populations, however, social relations are themselves very highly regulated through an often strictly formalized cultural grammar and syntax.

2.5 The Groupoid Free Energy

A formal equivalence class algebra can now be constructed by choosing different origin and end points $\mathbf{S}_0, \mathbf{S}_\infty$ and defining equivalence of two states by the existence of a high probability meaningful path connecting them with the same origin and end. Disjoint partition by equivalence class, analogous to orbit equivalence classes for dynamical systems, defines the vertices of the proposed network of cognitive dual languages, much enlarged beyond the spinglass example. We thus envision a network of metanetworks. Each vertex then represents a different equivalence class of information sources dual to a cognitive process. This is an abstract set of metanetwork "languages" dual to the cognitive processes of gene expression and development.

This structure generates a groupoid, in the sense of Weinstein (1996). States a_j, a_k in a set A are related by the groupoid morphism if and only if there exists a high probability grammatical path connecting them to the same base and end points, and tuning across the various possible ways in which that can happen—the different cognitive languages—parameterizes the set of equivalence relations and creates the (very large) groupoid. See the Mathematical Appendix for a summary of standard material on groupoids.

There is a hierarchy in groupoid structures. First, there is structure *within the system having the same base and end points*, as in Ciliberti et al. Second, there is a complicated groupoid structure defined by sets of dual information sources surrounding the variation of base and end points. We do not need to know what that structure is in any detail, but can show that its existence has profound implications.

First we examine the simple case, the set of dual information sources associated with a fixed pair of beginning and end states.

Taking the serial grammar/syntax model above, we find that not all high probability meaningful paths from $\mathbf{S}_0$ to $\mathbf{S}_\infty$ are the same. They are structured by the uncertainty of the associated dual information source, and that has a homological relation with free energy density.

Let us index possible dual information sources connecting base and end points by some set $A = \cup\alpha$. Argument by abduction from statistical physics is direct: Given metabolic energy density available at a rate M, and an allowed (fixed) characteristic development time τ, let $K = 1/\kappa M\tau$ for some appropriate scaling constant κ, so that $M\tau$ is total developmental free energy. Then we take the probability of a particular H_α as determined by a standard expression (e.g., Landau and Lifshitz 2007),

$$P[H_\beta] = \frac{\exp[-H_\beta K]}{\sum_\alpha \exp[-H_\alpha K]} \tag{2.10}$$

where the sum may, in fact, be a complicated abstract integral.

This is just a version of the fundamental probability relation from statistical mechanics, as above. The sum in the denominator, the partition function in statistical physics, is a crucial normalizing factor that allows the definition of $P[H_\beta]$ as a probability.

A basic requirement, then, is that the sum/integral always converges. K is the inverse product of a scaling factor, a metabolic energy density rate term, and a characteristic (presumed fixed) development time τ. The developmental energy might be raised to some power, e.g., $K = 1/(\kappa(M\tau)^b)$, suggesting the possibility of allometric scaling.

Some dual information sources will be "richer"/smarter than others, but, conversely, will require more metabolic energy for their completion.

While we might simply impose an equivalence class structure based on equal levels of energy/source uncertainty, producing a groupoid, we can do more by now allowing both source and end points to vary, as well as by imposing energy-level equivalence. This produces a far more highly structured groupoid that we now investigate.

Equivalence classes define groupoids, by standard mechanisms (e.g., Weinstein 1996; Brown 1987; Golubitsky and Stewart 2006). The basic equivalence classes, here involving both information source uncertainty level and the variation of $\mathbf{S}_0$ and $\mathbf{S}_\infty$, will define transitive groupoids, and higher order systems can be constructed by the union of transitive groupoids, having larger alphabets that allow more complicated statements in the sense of Ash above.

Again, given an appropriately scaled, dimensionless, fixed, inverse available metabolic energy density rate and development time, so that $K = 1/\kappa M\tau$, we propose that the metabolic-energy-constrained probability of an information source representing equivalence class G_i, H_{G_i}, will be given by the classic relation

$$P[H_{G_\alpha}] = \frac{\exp[-H_{G_\alpha}K]}{\sum_\beta \exp[-H_{G_\beta}K]}$$

where, now, we have shifted perspective, and *the sum/integral is over all possible elements of the largest available symmetry groupoid representing the equivalence class structure*. By the arguments of Ash above, compound sources, formed by the union of underlying transitive groupoids, being more complex, generally having richer alphabets, as it were, will all have higher free-energy-density-equivalents than those of the base (transitive) groupoids.

Let $Z_G \equiv \sum_\alpha \exp[-H_{G_\alpha}K]$. We now define the *Groupoid free energy* of the system, F_G, at inverse normalized metabolic energy density K, as

$$F_G[K] \equiv -\frac{1}{K}\log[Z_G[K]] \tag{2.11}$$

again following the standard arguments from statistical physics (again, Landau and Lifshitz 2007, or Feynman 2000).

Spontaneous Symmetry Breaking

The groupoid free energy permits introduction important ideas from statistical physics.

We have expressed the probability of an information source in terms of its relation to a fixed, scaled, available (inverse) metabolic free energy density, seen as a kind of equivalent (inverse) system temperature. This gives a statistical thermodynamic path leading to definition of a "higher" free energy construct—$F_G[K]$—to which we now apply Landau's fundamental heuristic phase transition argument (Landau and Lifshitz 2007; Skierski et al. 1989; Pettini 2007). See, in particular, Pettini (2007) for details.

Landau's insight was that second order phase transitions were usually in the context of a significant symmetry change in the physical states of a system, with one phase being far more symmetric than the other. A symmetry is lost in the transition, a phenomenon called spontaneous symmetry breaking, and symmetry changes are inherently punctuated. The greatest possible set of symmetries in a physical system is that of the Hamiltonian describing its energy states. Usually states accessible at lower temperatures will lack the symmetries available at higher temperatures, so that the lower temperature phase is less symmetric: The randomization of higher temperatures—in this case limited by available metabolic free energy densities—ensures that higher symmetry/energy states—mixed transitive groupoid structures—will then be accessible to the system. Absent high metabolic free energy rates and densities, however, only the simplest transitive groupoid structures can be manifest. A full treatment from this perspective seems to require invocation of groupoid representations, no small matter (e.g., Buneci 2003; Bos 2007).

Something like Pettini's (2007) Morse-Theory-based topological hypothesis can now be invoked, i.e., that changes in underlying groupoid structure are a necessary (but not sufficient) consequence of phase changes in $F_G[K]$. Necessity, but not sufficiency, is important, as it, in theory, allows mixed groupoid symmetries, leading to comorbidity in "condensation" dysfunctions.

Using this formulation, the mechanisms of epigenetic catalysis are accomplished by allowing the set B_1 above to span a distribution of possible "final" states $\mathbf{S}_\infty$. Then the groupoid arguments merely expand to permit traverse of both initial states and possible final sets, recognizing that there can now be a possible overlap in the latter, and the epigenetic effects are realized through the joint uncertainties $H(X_{G_\alpha}, Z)$, so that the epigenetic information source Z serves to direct as well the possible final states of X_{G_α}. Again, Scherrer and Jost (2007a,b) use information theory arguments to suggest something similar.

The Groupoid Atlas

The groupoid free energy inherently defines a groupoid atlas in the sense of Bak et al. (2006). Following closely Glazebrook and Wallace (2009a,b), the set of groupoids G_α comprise a groupoid atlas $\mathcal{A}$ as follows.

A family of local groupoids $(G_{\mathcal{A}})$ is defined with respective object sets $(X_{\mathcal{A}})_\alpha$, and a *coordinate system* $\Phi_{\mathcal{A}}$ of $\mathcal{A}$ equipped with a reflexive relation $\leq$. These satisfy the following conditions:

1. If $\alpha \leq \beta$ in $\Phi_{\mathcal{A}}$, then $(X_{\mathcal{A}})_\alpha \cap (X_{\mathcal{A}})_\beta$ is a union of components of $(G_{\mathcal{A}})$, that is, if $x \in (X_{\mathcal{A}})_\alpha \cap (X_{\mathcal{A}})_\beta$ and $g \in (G_{\mathcal{A}})_\alpha$ acts as $G : x \to y$, then $y \in (X_{\mathcal{A}})_\alpha \cap (X_{\mathcal{A}})_\beta$.
2. If $\alpha \leq \beta$ in $\Phi_{\mathcal{A}}$, then there is a groupoid morphism defined between the restrictions of the local groupoids to intersections

$$(G_{\mathcal{A}})_\alpha | (X_{\mathcal{A}})_\alpha \cap (X_{\mathcal{A}})_\beta \to (G_{\mathcal{A}})_\beta | (X_{\mathcal{A}})_\alpha \cap (X_{\mathcal{A}})_\beta,$$

and which is the identity morphism on objects.

Thus each of the G_α with its associated dual information source H_{G_α} constitutes a component of an atlas that incorporates the dynamics of an interactive system by means of the intrinsic groupoid actions.

These are matters currently under very active study (e.g., del Hoyo and Minian 2008).

2.6 "Phase Change" and the Developmental Holonomy Groupoid in Phenotype Space

There is a more direct way to look at phase transitions in cognitive, and here culturally driven, gene expression, adapting the topological perspectives of homotopy and holonomy directly within phenotype space.

We begin with ideas of directed homotopy.

In conventional topology one constructs equivalence classes of loops that can be continuously transformed into one another on a surface. The prospect of interest is to attempt to collapse such a family of loops to a point while remaining within the surface. If this cannot be done, there is a hole. Here we are concerned, as in Fig. 2.1, with sets of one-way developmental trajectories, beginning with an initial phenotype $\mathbf{S_i}$, and converging on some final phenotype, here characteristic (highly dynamic) brain phenotypes labeled, respectively, $\mathbf{S_n}$ and $\mathbf{S_o}$. One might view them as, respectively, "normal" and "other," and the developmental pathways as representing convergence on the two different configurations. The filled triangle represents the effect of a composite external epigenetic catalyst—including the effects of culture and culturally structured social interaction—acting at a critical developmental period represented by the initial phenotype $\mathbf{S_i}$.

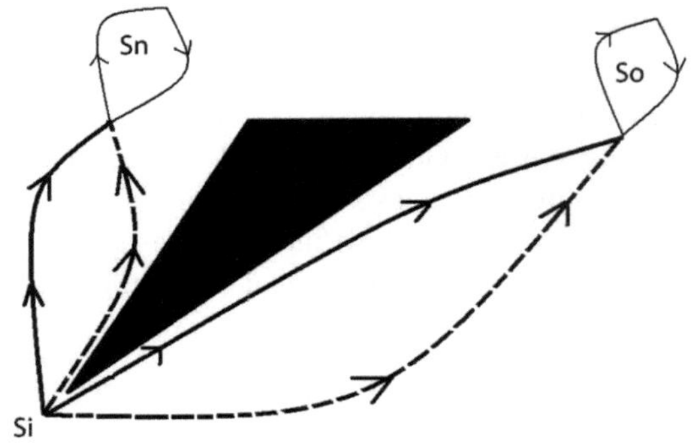

Fig. 2.1 Developmental homotopy equivalence classes in phenotype space. The set on one-way paths from $\mathbf{S_i}$ to $\mathbf{S_n}$ represents an equivalence class of developmental trajectories converging on a particular phenotype, here representing a highly dynamic normal mind structure. In the presence of a noxious external epigenetic catalyst, developmental trajectories can converge on an abnormal mind structure, represented by the dynamic phenotype $\mathbf{S_o}$

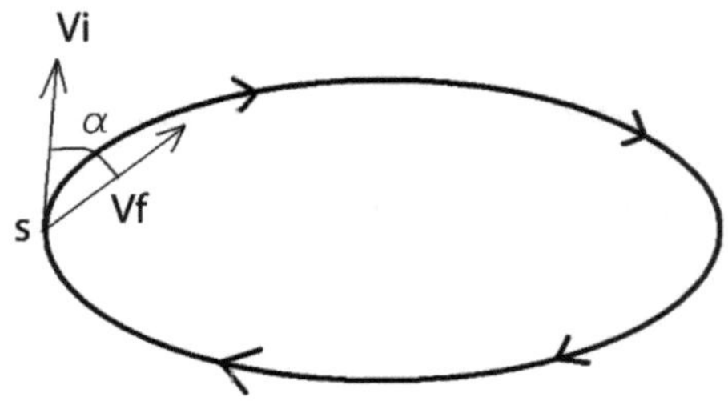

Fig. 2.2 Parallel transport of a tangent vector $V_i \rightarrow V_f$ around a loop on a manifold. Only for a geometrically flat object will the angle between the initial and final vectors be zero. By a fundamental theorem the path integral around the loop by parallel displacement is the surface integral of the curvature over the loop

We assume phenotype space to be directly measurable and to have a simple "natural" metric defining the difference between developmental paths.

Developmental paths continuously transformable into each other without crossing the filled triangle define equivalence classes characteristic of different information sources dual to cognitive gene expression, as above.

Given a metric on phenotype space, and given equivalence classes of developmental trajectories having more than one path each, we can *pair one-way developmental trajectories* to make loop structures. In Fig. 2.1 the solid and dotted lines above and below the filled triangle can be pasted together to make loops characteristic of the different developmental equivalence classes. Although Fig. 2.1 is represented as topologically flat, there is no inherent reason for the phenotype manifold itself to be flat. The existence of a metric in phenotype space permits determining the degree of curvature, using standard methods. Figure 2.2 shows a loop in phenotype space. Using the metric definition it is possible to *parallel transport* a tangent vector starting at point s around the loop, and to measure the angle between the initial and final vectors, as indicated. A central result from elementary metric geometry is that the angle α will be given by the integral of the curvature tensor of the metric over the interior of the loop (e.g., Frankel 2006, Sect. 9.6).

The *holonomy group* is defined as follows (e.g., Helgason 1962):

If s is a point on a manifold M having a metric, then the holonomy group of M is the group of all linear transformations of the tangent space M_s obtained by parallel translation along closed curves starting at s.

For Fig. 2.1 the *phenotype holonomy groupoid* is the disjoint union of the different holonomy groups corresponding to the different branches separated by "developmental shadows" induced by epigenetic information sources acting as developmental catalysts.

The relation between the phenotype groupoid as defined here and the phase transitions in $F_D[K]$ as defined above is an open question.

2.7 Holonomy on the Manifold of Dual Information Sources

Basic Structure

Glazebrook and Wallace (2009a) examined holonomy groupoid phase transition arguments for networks of interacting information sources dual to cognitive phenomena. A more elementary form of this arises directly through extending holonomy groupoid arguments to a manifold of different information source dual to cognitive phenomena as follows.

Different cognitive phenomena will have different dual information sources, and we are interested in the local properties of the system near a particular reference state. We impose a topology on the system, so that, near a particular "language" A, dual to an underlying cognitive process, there is an open set U of closely similar languages $\hat{A}$, such that $A, \hat{A} \subset U$. It may be necessary to coarse-grain the system's responses to define these information sources. The problem is to proceed in such a way as to preserve the underlying essential topology, while eliminating "high frequency noise." The formal tools for this can be found elsewhere, e.g., in Chap. 8 of Burago et al. (2001).

Since the information sources dual to the cognitive processes are similar, for all pairs of languages $A, \hat{A}$ in U, it is possible to:

1. Create an embedding alphabet which includes all symbols allowed to both of them.
2. Define an information-theoretic distortion measure in that extended, joint alphabet between any high probability (grammatical and syntactical) paths in A and $\hat{A}$, which we write as $d(Ax, \hat{A}x)$ (Cover and Thomas 2006). More detail on distortion measures is given in the section below on the Rate Distortion Theorem. Note that these languages do not interact, in this approximation.

3. Define a metric on U, for example,

$$\mathcal{M}(A,\hat{A}) = \left| \lim \frac{\int_{A,\hat{A}} d(Ax,\hat{A}x)}{\int_{A,A} d(Ax,A\hat{x})} - 1 \right| \tag{2.12}$$

integrating over the sets of high probability paths. Note that the integration in the denominator is over different paths within A itself, while in the numerator it is between different paths in A and $\hat{A}$. Other metric constructions on U seem possible, leading to similar results, just as different definitions of distortion lead to the same end in the Rate Distortion Theorem.

Structures weaker than a conventional metric would be of more general utility, but the mathematical complications are formidable.

Note that these conditions can be used to define equivalence classes of *languages* dual to cognitive processes, where previously we defined equivalence classes of *states* that could be linked by high probability, grammatical and syntactical paths connecting two phenotypes. This led to the characterization of different information sources. Here we construct an entity, formally a topological manifold, *that is an equivalence class of information sources*. This is, provided $\mathcal{M}$ is a conventional metric, a classic differentiable manifold. The set of such equivalence classes generates the *dynamical groupoid*, and questions arise regarding mechanisms, internal or external, which can break that groupoid symmetry.

Since H and $\mathcal{M}$ are both scalars, a "covariant" derivative can be defined directly as

$$dH/d\mathcal{M} = \lim_{\hat{A}\to A} \frac{H(A) - H(\hat{A})}{\mathcal{M}(A,\hat{A})} \tag{2.13}$$

where $H(A)$ is the source uncertainty of language A.

The essential point of a "covariant" derivative is that it is *independent of coordinate systems*, a condition this definition fulfills. As we will show below, this leads directly to ideas of a derivative along a tangent vector and to ideas of parallel transport leading to deep topological concepts such as holonomy. Introduction of a coordinate system in the definition of $\mathcal{M}$ quickly leads to the usual Christoffel symbols and the familiar geodesic equations (e.g., Wallace and Fullilove 2008, Sect. 8.3).

Suppose the system to be set in some reference configuration A_0.

To obtain the unperturbed dynamics of that state, impose a Legendre transform using this derivative, defining another scalar, an "entropy" analog, as

$$S \equiv H - \mathcal{M}dH/d\mathcal{M} \tag{2.14}$$

The simplest possible generalized Onsager relation—here seen as an empirical, fitted, equation like a regression model—is in terms of the gradient of S,

$$d\mathcal{M}/dt = LdS/d\mathcal{M} \quad (2.15)$$

where t is the time and $dS/d\mathcal{M}$ represents an analog to the thermodynamic force in a chemical system. This is seen as acting on the reference state A_0.

Again, explicit parameterization of $\mathcal{M}$—that is, imposing a coordinate system—introduces standard, and quite considerable, notational complications (Burago et al. 2001). Defining a metric for different cognitive dual languages parameterized by **K** leads to Riemannian, or even Finsler, geometries, including the usual geodesics (e.g., Wallace and Fullilove 2008; Glazebrook and Wallace 2009a,b).

The dynamics, as we have presented them so far, have been noiseless. The simplest generalized Onsager relation in the presence of noise might be rewritten as

$$d\mathcal{M}/dt = LdS/d\mathcal{M} + \sigma W(t)$$

where σ is a constant and $W(t)$ represents white noise. Again, S is seen as a function of the parameter $\mathcal{M}$. This leads directly to a family of classic stochastic differential equations of the form

$$d\mathcal{M}_t = L(t, \mathcal{M})dt + \sigma(t, \mathcal{M})dB_t \quad (2.16)$$

where L and σ are appropriately regular functions of t and $\mathcal{M}$, and dB_t represents the noise structure, characterized by its quadratic variation. In the sense of Emery (1989), this leads into complicated realms of stochastic differential geometry and related topics.

The natural generalization is to a system of developmental processes that interact via mutual information crosstalk, as described by Wallace and Wallace (2009).

"Coevolutionary" Development

The model can be applied to multiple interacting information sources representing simultaneous gene expression processes. This is, in a broad sense, a "coevolutionary" phenomenon in that the development of one process may affect that of others.

Most generally we assume that different cognitive developmental subprocesses of gene expression characterized by information sources H_m interact through chemical or other signals and assume that *different processes become each other's principal environments*, a broadly coevolutionary phenomenon.

We write

$$H_m = H_m(K_1 \ldots K_s, \ldots H_j \ldots) \quad (2.17)$$

where the K_s represent other relevant parameters and $j \neq m$.

The dynamics of such a system is driven by a recursive network of stochastic differential equations, similar to those used to study many other highly parallel dynamic structures (e.g., Zhu et al. 2007).

Letting the K_j and H_m all be represented as parameters Q_j (with the caveat that H_m not depend on itself), one can define, according to the generalized Onsager development of Wallace and Wallace (2009), "entropies" via the Legendre transforms as

$$S^m \equiv H_m - \sum_i Q_i \partial H_m / \partial Q_i$$

to obtain a complicated recursive system of phenomenological "Onsager relations" stochastic differential equations,

$$dQ_t^j = \sum_i [L_{j,i}(t, \ldots \partial S^m / \partial Q^i \ldots) dt + \sigma_{j,i}(t, \ldots \partial S^m / \partial Q^i \ldots) dB_t^i] \tag{2.18}$$

where, again for notational simplicity only, we have expressed both the H_j and the external K's in terms of the same symbols Q_j.

m ranges over the H_m and we could allow different kinds of "noise" dB_t^i, having particular forms of quadratic variation that may, in fact, represent a projection of environmental factors under something like a rate distortion manifold (Glazebrook and Wallace 2009).

It is important to realize that, for this formulation, one does not necessarily have the equivalent of "Onsager's fourth law" of thermodynamics, i.e., the symmetry relation $L_{i,j} = L_{j,i}$. This is because such a symmetry, at base, is a statement of local time reversal invariance (e.g., de Groot and Mazur 1984, pp. 35–41). But information sources are notoriously one-way in time, for example, someone speaking or writing in English is much more likely to utter the five-character string "the" than its reverse. Or, to put it another way, long palindromes, such as "Able was I ere I saw Elba," or "A man, a plan, a canal: Panama," are quite rare, and always relatively short, while information theory is based on asymptotic limit theorems most often involving very long strings of symbols.

As usual, for a system of equations like (2.18), there will be multiple quasi-stable nonequilibrium steady states, representing a class of generalized resilience modes accessible via holonomy punctuation.

There are, indeed, many possible patterns:

1. Setting the expectation of Eq. (2.18) equal to zero and solving for stationary points gives attractor states since the noise terms preclude unstable equilibria.
2. This system may, however, converge to limit cycle or "strange attractors" that are very highly dynamic.
3. What is converged to in both cases is not a simple state or limit cycle of states. Rather it is an equivalence class, or set of them, of generalized language information sources coupled by mutual interaction through crosstalk. Thus "stability" in this extended model represents particular patterns of ongoing dynamics rather than some identifiable "state," although such dynamics may be indexed by a "stable" set of phenotypes.

Here we become enmeshed in a system of highly recursive phenomenological stochastic differential equations, but at a deeper level than the standard stochastic chemical reaction model (e.g., Zhu et al. 2007), and in a dynamic rather than static manner: the objects of this system are equivalence classes of information sources and their crosstalk, rather than simple final states of a chemical system.

We have defined a groupoid for the system based on a particular set of equivalence classes of information sources dual to cognitive processes. That groupoid parsimoniously characterizes the available dynamical manifolds, and breaking of the groupoid symmetry by epigenetic crosstalk creates more complex objects of considerable interest. This leads to the possibility, indeed, the necessity of epigenetic *Deus ex Machina* mechanisms—analogous to programming, stochastic resonance, etc. —to force transitions between the different possible modes within and across dynamic manifolds. In one model the external "programmer" creates the manifold structure, and the system hunts within that structure for the "solution" to the problem according to equivalence classes of paths on the manifold. Noise, as with random mutation in evolutionary algorithms, precludes simple unstable equilibria, but not other possible structures.

Equivalence classes of *states* gave dual information sources. Equivalence classes of *information sources* give different characteristic dynamical manifolds. Equivalence classes of one-way developmental *paths* produce different directed homotopy topologies characterizing those dynamical manifolds. This introduces the possibility of having different quasi-stable modes *within* individual manifolds, and leads to ideas of holonomy and the holonomy groupoid of the set of quasi-stable developmental modes.

2.8 Rate Distortion Models

The Rate Distortion Theorem

The interaction between cognitive structures can be restated from a highly formal, but far more restricted, Rate Distortion Theorem perspective. Suppose a sequence of signals is generated by an information source dual to a cognitive process, Y having output $y^n = y_1, y_2, \ldots$. This is "digitized" in terms of the observed behavior of the system with which it communicates, say a sequence of observed behaviors $b^n = b_1, b_2, \ldots$. Often the b_i will happen in a characteristic "real time" τ. Assume each b^n is then deterministically retranslated back into a reproduction of the original biological signal,

$$b^n \rightarrow \hat{y}^n = \hat{y}_1, \hat{y}_2, \ldots .$$

Define a distortion measure $d(y, \hat{y})$ that compares the original to the retranslated path. Many such measures are possible. The Hamming distortion, for example, is

$$d(y, \hat{y}) = 1, y \neq \hat{y}$$

$$d(y, \hat{y}) = 0, y = \hat{y}$$

For continuous variates the squared error distortion is

$$d(y, \hat{y}) = (y - \hat{y})^2.$$

The distortion between *paths* y^n and $\hat{y}^n$ is defined as

$$d(y^n, \hat{y}^n) \equiv \frac{1}{n} \sum_{j=1}^{n} d(y_j, \hat{y}_j).$$

A remarkable fact of the Rate Distortion Theorem is that *the basic result is independent of the exact distortion measure chosen* (Cover and Thomas 2006; Dembo and Zeitouni 1998).

Suppose that with each path y^n and b^n-path retranslation into the y-language, denoted $\hat{y}^n$, there are associated individual, joint, and conditional probability distributions

$$p(y^n), p(\hat{y}^n), p(y^n, \hat{y}^n), p(y^n|\hat{y}^n).$$

The average distortion is defined as

$$D \equiv \sum_{y^n} p(y^n) d(y^n, \hat{y}^n) \tag{2.19}$$

It is possible, using the distributions given above, to define the information transmitted from the Y to the $\hat{Y}$ process using the Shannon source uncertainty of the strings:

$$I(Y, \hat{Y}) \equiv H(Y) - H(Y|\hat{Y}) = H(Y) + H(\hat{Y}) - H(Y, \hat{Y}) \tag{2.20}$$

where $H(\ldots, \ldots)$ is the joint and $H(\ldots | \ldots)$ the conditional uncertainty (Cover and Thomas 2006; Ash 1990).

If there is no uncertainty in Y given the retranslation $\hat{Y}$, then no information is lost, and the systems are in perfect synchrony.

In general, of course, this will not be true.

The *Rate Distortion Function* $R(D)$ for a source Y with a distortion measure $d(y, \hat{y})$ is defined as

$$R(D) = \min_{p(y,\hat{y}); \sum_{(y,\hat{y})} p(y) p(y|\hat{y}) d(y,\hat{y}) \leq D} I(Y, \hat{Y}) \tag{2.21}$$

The minimization is over all conditional distributions $p(y|\hat{y})$ for which the joint distribution $p(y, \hat{y}) = p(y)p(y|\hat{y})$ satisfies the average distortion constraint (i.e., average distortion $\leq D$).

The *Rate Distortion Theorem* states that $R(D)$ is the minimum necessary rate of information transmission which ensures the communication between the modules does not exceed average distortion D. Thus $R(D)$ defines a minimum necessary channel capacity. Cover and Thomas (2006) or Dembo and Zeitouni (1998) provide details. The Rate Distortion Function has been calculated for a number of systems using Lagrange multiplier and related methods.

There is an absolutely central fact characterizing the Rate Distortion Function: Cover and Thomas (2006) show that *$R(D)$ is necessarily a decreasing convex function of D* for any reasonable definition of distortion.

That is, *$R(D)$ is always* a reverse J-shaped curve. This will prove crucial for the overall argument. Indeed, convexity is an exceedingly powerful mathematical condition, and permits deep inference (e.g., Rockafellar 1970). Ellis (1985, Chap. VI) applies convexity theory to conventional statistical mechanics.

For a Gaussian channel having noise with zero mean and variance σ^2 under the squared distortion measure (Cover and Thomas 2006),

$$R(D) = 1/2 \log[\sigma^2/D], 0 \leq D \leq \sigma^2$$
$$R(D) = 0, D \geq \sigma^2 \quad (2.22)$$

Recall, now, the relation between information source uncertainty and channel capacity (e.g., Ash 1990):

$$H[\mathbf{X}] \leq C \quad (2.23)$$

where H is the uncertainty of the source X and C the channel capacity, defined according to the relation (Ash 1990)

$$C \equiv \max_{P(X)} I(X|Y) \quad (2.24)$$

where $P(X)$ is chosen so as to maximize the rate of information transmission along a channel Y.

Finally, recall the analogous definition of the Rate Distortion Function from Eq. (2.21), again an extremum over a probability distribution.

Rate Distortion Dynamics

$R(D)$ defines the minimum channel capacity necessary for the system to have average distortion less than or equal to D, placing a limit on information source uncertainty. Thus, we suggest distortion measures can drive information system dynamics. That is, the Rate Distortion Function also has a homological relation to free energy density, similar to the relation between free energy density and information source uncertainty.

We are led to propose, as a heuristic, that the dynamics of cognitive modules interacting in a characteristic "real time" τ will be constrained by the system as described in terms of a parameterized Rate Distortion Function. To do this, take R as parameterized, not only by the distortion D, but also by some vector of variates $\mathbf{Q} = (Q_1, \ldots, Q_k)$, for which the first component is the average distortion. The assumed dynamics are, as in Wallace and Wallace (2008), then driven by gradients in the rate distortion disorder defined as the Legendre transform of the Rate Distortion Function

$$S_R \equiv R(\mathbf{Q}) - \sum_{i=1}^{k} Q_i \partial R / \partial Q_i \tag{2.25}$$

This leads to the deterministic and stochastic systems of equations analogous to the Onsager relations of nonequilibrium thermodynamics:

$$dQ_j/dt = \sum_i L_{j,i} \partial S_R / \partial Q_i \tag{2.26}$$

and

$$dQ_t^j = L^j(Q_1, \ldots, Q_k, t)dt + \sum_i \sigma^{j,i}(Q_1, \ldots, Q_k, t)dB_t^i \tag{2.27}$$

where the dB_t^i represent added, often highly structured, stochastic "noise" whose properties are characterized by the quadratic variation (e.g., Protter 1995).

Even for this simplified structure, it is not clear under what circumstances "Onsager reciprocal relations" are possible. Since average distortion is a scalar, however, some systems may indeed display the kind of time reversal invariance required for those symmetries.

A central focus of this chapter, however, is to generalize these equations in the face of richer structures, for example, interactions between cognitive modules that may not be time-reversible, the existence of characteristic time constants within nested processes, and the influence of an embedding source of free energy.

For a simple Gaussian channel with noise having zero mean and variance σ^2, an entropy can be defined as the Legendre transform

$$S_R(D) = R(D) - DdR(D)/dD = 1/2 \log(\sigma^2/D) + 1/2 \tag{2.28}$$

The simplest possible Onsager relation becomes

$$dD/dt = -\mu dS_R/dD = \frac{\mu}{2D} \tag{2.29}$$

where $-dS_R/dD$ represents the force of an entropic wind, a kind of internal dissipation inevitably driving the real time, system of interacting (cognitive) information sources toward greater distortion.

This has the solution

$$D = \sqrt{\mu t} \tag{2.30}$$

so that the average distortion increases monotonically with time, for this model.

A central observation is that *similar results must necessarily apply to any of the reverse-J-shaped relations that inevitably characterize* $R(D)$, since the Rate Distortion Function is necessarily a convex decreasing function of the average distortion D, whatever distortion measure is chosen. Again, see Cover and Thomas (2006) for details.

The explicit implication is that a system of cognitive modules interacting in real time will inevitably be subject to a relentless entropic force, requiring a constant free energy expenditure for maintenance of some fixed average distortion in the communication between them: The distortion in the communication between two interacting modules will, without free energy input, have time dependence

$$D = f(t) \tag{2.31}$$

with $f(t)$ monotonic increasing in t.

This necessarily leads to the punctuated failure of the system.

Note that Eq. (2.30) is similar to classical Brownian motion as treated by Einstein: Let $p(x, t)dx$ be the probability a particle located at the origin at time zero and undergoing Brownian motion is found at locations $x \to x + dx$ at time t. Then, p satisfies the diffusion equation $\partial p(x, t)/\partial t = \mu \partial^2 p(x, t)/\partial x^2$. Einstein's solution is that

$$p(x, t) = \frac{1}{\sqrt{4\pi \mu t}} \exp[-x^2/4\mu t].$$

It is easy to show that the standard deviation of the particle position increases in proportion to $\sqrt{\mu t}$, just as above.

Some comment is appropriate. Following, e.g., Chung and Williams (1990), a process $B = B_t, t \in \mathcal{R}_+$ is called a Brownian motion in $\mathcal{R}_+$ iff:

1. for $0 \le s < t < \infty$, $B_t - B_s$ is a normally distributed random variate with mean zero and variance $|t - s|$.
2. for $0 \le t_0 < t_1 < \cdots < t_k < \infty$,

$$\{B_{t_0}; B_{t_j} - B_{t_{j-1}}, j = 1, \ldots, k\}$$

is a set of independent random variates.

An information source, of course, generates a *highly correlated sequence* that grossly violates these simple assumptions. What we have shown is that the *distortion* in the communication between interacting cognitive modules, under appropriate empirical Onsager relations, can behave as if it were undergoing Brownian motion. This is a simple, but far from trivial, result.

Prandolini and Moody (1995) have, in fact, observed something much like this in the time base error of recorded signals. Wow and flutter are the instantaneous speed error between recording and reproduction epochs. The time base error (TBE) in the reproduced signal is a function of the wow and flutter. They show, empirically, that the nonperiodic TBE is a *fractional Brownian motion*. The implication is that the nonperiodic flutter is fractional Gaussian, and thus what they call a "blind" TBE system is impractical for the design of a TBE compensation system.

Normalized fractional Brownian motion on $(0, t), t \in \mathcal{R}_+$ is a continuous time Gaussian process starting at zero, with mean zero, and having the covariance function (Beran 1994)

$$E[B^H(t)B^H(s)] = (1/2)[|t|^{2H} + |s|^{2H} - |t-s|^{2H}].$$

If $H = 1/2$ the process is a regular Brownian motion. Otherwise, for $H > 1/2$, the increments are positively correlated, and for $H < 1/2$, negatively correlated.

We will explore this kind of relation in more detail below.

Rate Distortion Coevolutionary Dynamics

A simplified version of Eq. (2.18) can be constructed using the Rate Distortion Functions for mutual crosstalk between a set of interacting cognitive modules, using the homology of the Rate Distortion Function itself with free energy, as driven by the inherent convexity of the Rate Distortion Function *R*(*D*). That convexity is, in fact, why we invoke the Rate Distortion Function.

Given different cognitive processes $1 \ldots s$, the quantities of special interest thus become the mutual Rate Distortion Functions $R_{i,j}$ characterizing communication (and the distortion $D_{i,j}$) between them, while the essential parameters remain the characteristic time constants of each process, $\tau_j, j = 1 \ldots s$, and an overall, embedding, available free energy density, F.

Taking the Q^α to run over all the relevant parameters and mutual Rate Distortion Functions (including distortion measures $D_{i,j}$), Eq. (2.14) becomes

$$S_R^{i,j} \equiv R_{i,j} - \sum_k Q^k \partial R_{i,j} / \partial Q^k \quad (2.32)$$

Equation (2.18) accordingly becomes

$$dQ_t^\alpha = \sum_{\beta=(i,j)} [L_\beta(t, \ldots \partial S_R^\beta / \partial Q^\alpha \ldots)dt + \sigma_\beta(t, \ldots \partial S_R^\beta / \partial Q^\alpha \ldots)dB_t^\beta] \quad (2.33)$$

and this generalizes the treatment in terms of crosstalk, its distortion, the inherent time constants of the different cognitive modules, and the overall available free energy density.

This is a very complicated structure indeed, but its general dynamical behaviors will obviously be analogous to those described just above. For example, setting the expectation of Eq. (2.33) to zero gives the "coevolutionary stable states" of a system of interacting cognitive modules. Again, limit cycles and strange attractors seem possible as well. And again, what is converged to is a dynamic behavior pattern, not some fixed "state." And again, such a system will display highly punctuated dynamics almost exactly akin to resilience domain shifts in ecosystems (e.g., Holling 1973, 1992; Gunderson 2000). Indeed, the formalism seems directly applicable to ecosystem studies.

And again, because these are highly self-dynamic cognitive phenomena and not simple crystals or other physical objects, it may not often be possible to invoke time reversal invariance to give Onsager-like reciprocal symmetries to Eq. (2.33).

2.9 Expanding the Mathematical Approach

We have, in the context of a tunable epigenetic catalysis, developed several phase transition/branching models of cognitive gene expression based on groupoid structures that may be applied to the development of the human mind and its dysfunctions, as known to be particularly influenced by embedding culture. The first used Landau's spontaneous symmetry breaking to explore phase transitions in a groupoid free energy $F_D[K]$. The second examined a holonomy groupoid in phenotype space generated by disjoint developmental homotopy equivalence classes, and "loops" constructed by pairing one-way development paths. The third introduced a metric on a manifold of different information sources dual to cognitive gene expression, leading to a more conventional picture of parallel transport around a loop leading to holonomy. The dynamical groupoid of Wallace and Fullilove (2008, Sect. 3.8) is seen as involving a disjoint union across underlying manifolds that produces a holonomy groupoid in a natural manner.

There are a number of outstanding mathematical questions.

The first is the relation between the Landau formalism and the structures of phenotype space S and those of the associated manifold of dual information sources, the manifold M having metric $\mathcal{M}$. How does epigenetic catalysis in M-space imposes structure on S-space? How is this related to spontaneous symmetry breaking?

What would a stochastic version of the theory, in the sense of Emery (1989), look like? It is quite possible, using appropriate averages of the stochastic differential equations that arise naturally, to define parallel transport, holonomy, and the like for these structures. In particular a stochastic extension of the results of the first question would seem both fairly direct and interesting from a real-world perspective, as development is always "noisy."

The construction of loops from directed homotopy arcs in Fig. 2.1 is complicated by the necessity of imposing a consistent piecewise patching rule for parallel translation at the end of each arc, say from $\mathbf{S}_i$ to $\mathbf{S}_n$. This can probably be done by some standard fiat, but the details will likely be messy.

On another matter, we have imposed metrics on S and M space, making possible a fairly standard manifold analysis of complex cognitive processes of gene expression and development. While this is no small thing, the "natural" generalization, given the ubiquity of groupoids across our formalism, would be to a more complete groupoid atlas treatment in the spirit of section "The Groupoid Atlas". The groupoid atlas permits a weaker structure compared with that of a conventional manifold since no condition of compatibility between arbitrary overlaps of the patches is necessary. It is possible that the groupoid atlas will become, to complicated problems in biological cognitive process, something of what the Riemannian manifold has been to physics.

With regard to questions of "smoothness," we are assuming that the cognitive landscape of gene expression is sufficiently rich that discrete paths can be well approximated as continuous where necessary, the usual physicist's hack.

Finally, Sects. 2.6 and 2.7 are based on existence of more-or-less conventional metrics, and this may not be a good approximation to many real systems. Extending topological phase transition theory to "weaker" topologies, e.g., Finsler geometries and the like, is not a trivial task.

2.10 Discussion

Culturally structured psychosocial stress, and similar noxious exposures, can write distorted images of themselves onto human ontology—both child growth, and, if sufficiently powerful, adult development as well—by a variety of mechanisms, initiating a punctuated trajectory to characteristic forms of comorbid mind/body dysfunction. This occurs in a manner recognizably analogous to resilience domain shifts affecting stressed ecosystems (e.g., Wallace 2008; Holling 1973; Gunderson 2000). Consequently, like ecosystem restoration, reversal or palliation may often be exceedingly difficult once a generalized domain shift has taken place. Thus a public health approach to the prevention of mental disorders may be paramount: rather than seeking to understand why half a population does not respond to the LD50 of a teratogenic environmental exposure, one seeks policies and social reforms that limit the exposure.

Both sociocultural and epigenetic environmental influences—like gene methylation—are heritable, in addition to genetic mechanisms. The missing heritability of complex diseases that Manolio et al. (2009) seek to find in more and better gene studies is most likely dispersed within the "dark matter" of these two other systems of heritage that together constitute the larger, and likely highly synergistic, regulatory machinery for gene expression. More and more purely genetic studies would, under such circumstances, be akin to using increasingly powerful microscopes to look for cosmic membranes of strewn galaxies.

A crucial matter is the conversion of the probability models we present here into statistical tools suitable for analyzing real data, and hence actually testing the theoretical models we present here. Some work in this direction has been done in Sect. 2.8, but the problem involves not just programming such models for use, but identifying appropriate real-world problems, assembling available data sets, transforming the data as needed for the models, and actually applying the statistical models. Indeed, the environmental health literature contains numerous examples of developmental deviations due to either chemical exposures or interaction between chemical and socioeconomic exposures, and these could serve as sources of data for direct analysis (e.g., Needleman et al. 1996; Fullilove 2004; Dietrich et al. 2001; Miranda et al. 2007; Glass et al. 2009; Jacobson and Jacobson (2002); Shankardass et al. 2009; Clougherty et al. 2007; Ben-Jonathan et al. 2009; Karp et al. 2005; Sarlio-Lahteenkorva and Lahelma 2001; Wallace and Wallace 2005; Wallace et al. 2003). Thus, quite a number of data sets exist in the environmental health and socioeconomic epidemiological literature that could be subjected to meta-analysis and other review for model verification and fitting. Our topological models, when converted to statistical tools for data analysis, hold great potential for understanding developmental trajectories and interfering factors (teratogens) through the life course. Sets of cross cultural variants of these data focusing specifically on mental disorders would be needed to address the particular concerns of this chapter.

Nonetheless, what we have done is of no small interest for understanding the ontology of the human mind and its pathologies. West-Eberhard (2003, 2005) argues that any new input, whether it comes from the genome, like a mutation, or from the external environment, like a temperature change, a pathogen, or a parental opinion, has a developmental effect only if the preexisting phenotype is responsive to it. A new input causes a reorganization of the phenotype, or "developmental recombination." In developmental recombination, phenotypic traits are expressed in new or distinctive combinations during ontogeny, or undergo correlated quantitative change in dimensions. Developmental recombination can result in evolutionary divergence at all levels of organization.

According to West-Eberhard, individual development can be visualized as a series of branching pathways. Each branch point is a developmental decision, or switch point, governed by some regulatory apparatus, and each switch point defines a modular trait. Developmental recombination implies the origin or deletion of a branch and a new or lost modular trait. The novel regulatory response and the novel trait originate simultaneously. Their origins are, in fact, inseparable events: There cannot, West-Eberhard concludes, be a change in the phenotype, a novel phenotypic state, without an altered developmental pathway.

Our analysis provides a new formal picture of this process as it applies to human development: The normal branching of developmental trajectories, and the disruptive impacts of teratogenic events of various kinds, can be described in terms of a growing sequence of holonomy groupoids, each associated with a set of dual information sources representing patterns of cognitive gene expression catalyzed by epigenetic information sources that, for humans, must include culture and culturally modulated social interaction as well as more direct mechanisms like gene methylation. This is a novel way of looking at human development and its disorders that may prove to be of some use. The most important innovation of this work, however, seems to be the natural incorporation of embedding culture as an essential component of the epigenetic regulation of human ontology, and in the effects of environment on the expression of a broad spectrum of developmental disorders: the missing heritability of complex diseases found.

Acknowledgements We thank M. Weissman for critical comments useful in revision, and S. Heine for access to the Henrich et al. and H. Kim preprints. The opinions expressed, however, remain distinctly those of the authors.

References

Arnett, J. 2008. The neglected 95%. *The American Psychologist* 63: 602–614.
Ash, R. 1990. *Information Theory*. New York: Dover.
Atlan, H., and I. Cohen. 1998. Immune information, self-organization, and meaning. *International Immunology* 10: 711–717.
Atmanspacher, H. 2006. Toward an information theoretical implementation of contextual conditions for consciousness. *Acta Biotheoretica* 54: 157–160.
Backdahl, L., A. Bushell, and S. Beck. 2009. Inflammatory signalling as mediator of epigenetic modulation in tissue-specific chronic inflammation. *The International Journal of Biochemistry and Cell Biology*. doi:10.1016/j.biocel.2008.08.023.
Bak, A., R. Brown, G. Minian, and T. Porter. 2006. Global actions, groupoid atlases and related topics. *Journal of Homotopy and Related Structures* 1: 1–54.
Bebbington, P. 1993. Transcultural aspects of affective disorders. *International Review of Psychiatry* 5: 145–156.
Ben-Jonathan, N., E. Hugo, and T. Brandenbourg. 2009. Effects of bisphenol A on adipokine release from human adipose tissue: Implications for the metabolic syndrome. *Molecular Cell Endocrinology* 304: 49–54.
Bennett, C. 1988. Logical depth and physical complexity. In *The Universal Turing Machine: A Half-Century Survey*, ed. R. Herkin, 227–257. Oxford: Oxford University Press.
Beran, J. 1994. *Statistics for Long-Memory Processes*. New York: Chapman and Hall.
Bos, R. 2007. Continuous representations of groupoids. arXiv:math/0612639.
Brown, R. 1987. From groups to groupoids: a brief survey. *Bulletin of the London Mathematical Society* 19: 113–134.
Brown, G., T. Harris, and J. Peto. 1973. Life events and psychiatric disorders, II: Nature of causal link. *Psychological Medicine* 3: 159–176.
Buneci, M. 2003. *Representare de Groupoizi*. Timisoara: Editura Mirton.
Burago, D., Y. Burago, and S. Ivanov. 2001. *A Course in Metric Geometry*. Graduate Studies in Mathematics, vol. 33. Providence, RI: American Mathematical Society.

Chung, K., and R. Williams. 1990. *Introduction to Stochastic Integration*, 2nd ed. Boston, MA: Birkhauser.
Ciliberti, S., O. Martin, and A. Wagner. 2007a. Robustness can evolve gradually in complex regulatory networks with varying topology. *PLoS Computational Biology* 3(2): e15.
Ciliberti, S., O. Martin, and A. Wagner. 2007b. Innovation and robustness in complex regulatory gene networks. *Proceeding of the National Academy of Sciences* 104: 13591–13596.
Clougherty, J., J. Levy, L. Kubzansky, P. Ryan, S. Suglia, M. Canner, and R. Wright. 2007. Synergistic effects of traffic-related air pollution and exposure to violence on urban asthma etiology. *Environmental Health Perspectives* 115: 1140–1146.
Cohen, I. 2006. Immune system computation and the immunological homunculus. In *MoDELS 2006*, ed. O. Nierstrasz, J. Whittle, D. Harel, and G. Reggio. Lecture Notes in Computer Science, vol. 4199, 499–512. Heidelberg: Springer.
Cohen, I., and D. Harel. 2007. Explaining a complex living system: Dynamics, multi-scaling, and emergence. *Journal of the Royal Society: Interface* 4: 175–182.
Cover, T., and J. Thomas. 2006. *Elements of Information Theory*, 2nd ed. New York: Wiley.
de Groot, S., and R. Mazur. 1984. *Non-equilibrium Thermodynamics*. New York: Dover.
Dehaene, S., and L. Naccache. 2001. Towards a cognitive neuroscience of consciousness: Basic evidence and a workspace framework. *Cognition* 79: 1–37.
del Hoyo, M., and E. Minian. 2008. Classical invariants for global actions and groupoid atlases. *Applied Categorical Structures* 18: 689–721.
Dembo, A., and O. Zeitouni. 1998. *Large Deviations: Techniques and Applications*, 2nd ed. New York: Springer.
Dietrich, K., R. Douglas, P. Succop, O. Berger, and R. Bornschein. 2001. Early exposure to lead and juvenile delinquency. *Neurotoxicology and Teratology* 23: 511–518.
Dohrenwend, B.P., and B.S. Dohrenwend. 1974. Social and cultural influences on psychopathology. *Annual Review of Psychology* 25: 417–452.
Dretske, F. 1994. The explanatory role of information. *Philosophical Transactions of the Royal Society A* 349: 59–70.
Durham, W. 1991. *Coevolution: Genes, Culture and Human Diversity*. Palo Alto, CA: Stanford University Press.
Eaton, W. 1978. Life events, social supports, and psychiatric symptoms: A re-analysis of the New Haven data. *Journal of Health and Social Behavior* 19: 230–234.
Ellis, R. 1985. *Entropy, Large Deviations, and Statistical Mechanics*. New York: Springer.
Emery, M. 1989. *Stochastic Calculus on Manifolds*. New York: Springer.
Feynman, R. 2000. *Lectures on Computation*. New York: Westview Press.
Foley, D., J. Craid, R. Morley, C. Olsson, T. Dwyer, K. Smith, and R. Saffery. 2009. Prospects for epigenetic epidemiology. *American Journal of Epidemiology* 169: 389–400.
Frankel, T. 2006. *The Geometry of Physics: An Introduction*, 2nd ed. Cambridge: Cambridge University Press.
Fullilove, M. 2004. *Root Shock: How Tearing Up City Neighborhoods Hurts America and What We Can Do About It*. New York: Ballantine Books.
Gilbert, P. 2001. Evolutionary approaches to psychopathology: The role of natural defenses. *Australian and New Zealand Journal of Psychiatry* 35: 17–27.
Glass, T., K. Bandeen-Roche, M. McAtee, K. Bolla, A. Todd, and B. Schwartz. 2009. Neighborhood psychosocial hazards and the association of cumulative lead dose with cognitive function in older adults. *American Journal of Epidemiology* 169: 683–692.
Glazebrook, J.F., and R. Wallace, 2009, Rate distortion manifolds as model spaces for cognitive information. *Informatica* 33: 309–346.
Glazebrook, J.F., and R. Wallace. 2009a. Small worlds and red queens in the global workspace: An information-theoretic approach. *Cognitive Systems Research* 10: 333–365.
Glazebrook, J.F., and R. Wallace. 2009b. Rate distortion manifolds as model spaces for cognitive information. *Informatica* 33: 309–346.
Golubitsky, M., and I. Stewart. 2006. Nonlinear dynamics and networks: The groupoid formalism. *Bulletin of the American Mathematical Society* 43: 305–364.

Gunderson, L. 2000. Ecological resilience – In theory and application. *Annual Reviews of Ecological Systematics* 31: 425–439.
Heine, S. 2001. Self as cultural product: An examination of East Asian and North American selves. *Journal of Personality* 69: 881–906.
Helgason, S. 1962. *Differential Geometry and Symmetric Spaces*. New York: Academic.
Henrich, J., S. Heine, and A. Norenzayan. 2010. The weirdest people in the world? *Behavioral and Brain Sciences* 33: 61–83.
Holling, C. 1973. Resilience and stability of ecological systems. *Annual Reviews of Ecological Systematics* 4: 1–23.
Holling, C. 1992. Cross-scale morphology, geometry, and dynamics of ecosystems. *Ecological Monographs* 62: 447–502.
Hoppensteadt, F., and E. Izhikevich. 1997. *Weakly Connected Neural Networks*. New York: Springer.
Jablonka, E. 2004. Epigenetic epidemiology. *International Journal of Epidemiology* 33: 929–935.
Jablonka, E., and M. Lamb. 1995. *Epigenetic Inheritance and Evolution: The Lamarckian Dimension*. Oxford: Oxford University Press.
Jablonka, E., and M. Lamb. 1998. Epigenetic inheritance in evolution. *Journal of Evolutionary Biology* 11: 159–183.
Jacobson, J., and S. Jacobson. 2002. Breast-feeding and gender as moderators of teratogenic effects on cognitive development. *Neurotoxicological Teratology* 24: 349–358.
Jaenisch, R., and A. Bird. 2003. Epigenetic regulation of gene expression: How the genome integrates intrinsic and environmental signals. *Nature Genetics Supplement* 33: 245–254.
Jenkins, J., A. Kleinman, and B. Good. 1990. Cross-cultural studies of depression. In *Advances in Mood Disorders: Theory and Research*, ed. J. Becker and A. Kleinman, 67–99. Los Angeles, CA: L. Erlbaum.
Johnson-Laird, P., F. Mancini, and A. Gangemi. 2006. A hyper-emotion theory of psychological illnesses. *Psychological Reviews* 113: 822–841.
Karp, R., C. Chen, and A. Meyers. 2005. The appearance of discretionary income: Influence on the prevalence of under- and over-nutrition. *International Journal of Equity in Health* 4: 10.
Khinchin, A. 1957. *Mathematical Foundations of Information Theory*. New York: Dover.
Kleinman, A., and A. Cohen. 1997. Psychiatry's global challenge. *Scientific American* 276(3): 86–89.
Kleinman, A., and B. Good. 1985. *Culture and Depression: Studies in the Anthropology of Cross-Cultural Psychiatry of Affect and Depression*. Berkeley, CA: University of California Press.
Landau, L., and E. Lifshitz. 2007. *Statistical Physics*, 3rd ed., Part I. New York: Elsevier.
Manolio, T., F. Collins, N. Cox, et al. 2009. Finding the missing heritability of complex diseases. *Nature* 461: 747–753.
Manson, S. 1995. Culture and major depression: Current challenges in the diagnosis of mood disorders. *Psychiatric Clinics of North America* 18: 487–501.
Markus, H., and S. Kitayama. 1991. Culture and the self- implications for cognition, emotion, and motivation. *Psychological Review* 98: 224–253.
Marsella, A. 2003. Cultural aspects of depressive experience and disorders. In *Online Readings in Psychology and Culture (Unit 9, Chapter 4)*, ed. W. Lonner, D. Dinnel, S. Hays, and D. Sattler. Bellingham, WA: Center for Cross-Cultural Research, Western Washington University. http://www.wwu.edu/~culture
Matsuda, T., and R. Nisbett. 2006. Culture and change blindness. *Cognitive Science: A Multidisciplinary Journal* 30: 381–399.
Maturana, H., and F. Varela. 1980. *Autopoiesis and Cognition*. Dordrecht: Reidel.
Maturana, H., and F. Varela. 1992. *The Tree of Knowledge*. Boston, MA: Shambhala Publications.
Miranda, M., D. Kim, M. Overstreet Galeano, C. Paul, A. Hull, and S. Morgan. 2007. The relationship between early childhood blood lead levels and performance on end-of-grade tests. *Environmental Health Perspectives* 115: 1242–1247.
Needleman, H., J. Riess, M. Tobin, G. Biesecker, and J. Greenhouse. 1996. Bone lead levels and delinquent behavior. *Journal of the American Medical Association* 275: 363–369.

Nesse, R. 2000. Is depression an adaptation? *Archives of General Psychiatry* 57: 14–20.

Nisbett, R., K. Peng, C. Incheol, and A. Norenzayan. 2001. Culture and systems of thought: Holistic vs. analytic cognition. *Psychological Review* 108: 291–310.

Norenzayan, A., and S. Heine. 2005. Psychological universals: What are they and how can we know? *Psychological Bulletin* 131: 763–784.

O'Nuallain, S. 2008. Code and context in gene expression, cognition, and consciousness. In *The Codes of Life: The Rules of Macroevolution*, ed. M. Barbiere, Chapter 15, 347–356. New York: Springer.

Pettini, M. 2007. *Geometry and Topology in Hamiltonian Dynamics and Statistical Mechanics*. New York: Springer.

Prandolini, R., and M. Moody. 1995. Brownian nature of the Time-Base Error in tape recordings. *Journal of the Audio Engineering Society* 43: 241–247.

Protter, P. 1995. *Stochastic Integration and Differential Equations: A New Approach*. New York: Springer.

Richerson, P., and R. Boyd. 2004. *Not by Genes Alone: How Culture Transformed Human Evolution*. Chicago, IL: Chicago University Press.

Risch, N., R. Herrell, T. Lehner, K. Liang, L. Eaes, J. Hoh, A. Griem, M. Kovacs, J. Ott, and K. Merikangas. 2009. Interaction between the serotonin transporter gene (5-HTTLPR), stressful life events, and risk of depression. *Journal of the American Medical Association*, 301: 2462–2472.

Rockafellar, R. 1970. *Complex Analysis*. Princeton, NJ: Princeton University Press.

Sarlio-Lahteenkorva, S., and E. Lahelma. 2001. Food insecurity is associated with past and present economic disadvantage and body mass index. *Journal of Nutrition* 131: 2880–2884.

Scherrer, K., and J. Jost. 2007a. The gene and the genon concept: A functional and information-theoretic analysis. *Molecular Systems Biology* 3: 87–93.

Scherrer, K., and J. Jost. 2007b. Gene and genon concept: Coding versus regulation. *Theory in Bioscience* 126: 65–113.

Shankardass, K., McConnell, R., Jerrett, M., Milam, J., Richardson, J., and K. Berhane. 2009. Parental stress increases the effect of traffic-related air pollution on childhood asthma incidence. *Proceedings of the National Academy of Sciences*, 106: 12406–12411.

Skierski, M., A. Grundland, and J. Tuszynski. 1989. Analysis of the three-dimensional time-dependent Landau-Ginzburg equation and its solutions. *Journal of Physics A (Math. Gen.)* 22: 3789–3808.

Toulouse, G., S. Dehaene, and J. Changeux. 1986. Spin glass model of learning by selection. *Proceedings of the National Academy of Sciences* 83: 1695–1698.

Turner, B. 2000. Histone acetylation and an epigenetics code. *Bioessays* 22: 836–845.

Wallace, R. 2005. *Consciousness: A Mathematical Treatment of the Global Neuronal Workspace Model*. New York: Springer.

Wallace, R. 2007. Culture and inattentional blindness. *Journal of Theoretical Biology* 245: 378–390.

Wallace, R. 2008. Developmental disorders as pathological resilience domains. *Ecology and Society* 13: 29 (online).

Wallace, R., and M. Fullilove. 2008. *Collective Consciousness and its Discontents: Institutional Distributed Cognition, Racial Policy, and Public Health in the United States*. New York: Springer.

Wallace, D., and R. Wallace. 1998. Scales of geography, time, and population: The study of violence as a public health problem. *American Journal of Public Health* 88: 1853–1858.

Wallace, R., and D. Wallace. 2005. Structured psychosocial stress and the US obesity epidemic. *Journal of Biological Systems* 13: 363–384.

Wallace, R., and D. Wallace. 2008. Punctuated equilibrium in statistical models of generalized coevolutionary resilience: How sudden ecosystem transitions can entrain both phenotype expression and Darwinian selection. *Transactions on Computational Systems Biology IX, LNBI* 5121: 23–85.

Wallace, R., and D. Wallace. 2009. Code, context, and epigenetic catalysis in gene expression. *Transactions on Computational Systems Biology XI, LNBI* 5750: 283–334.

Wallace, D., R. Wallace, and V. Rauh. 2003. Community stress, demoralization, and body mass index: Evidence for social signal transduction. *Social Science and Medicine* 56: 2467–2478.

Weinstein, A. 1996. Groupoids: unifying internal and external symmetry. *Notices of the American Mathematical Association* 43: 744–752.

West-Eberhard, M. 2003. *Developmental Plasticity and Evolution*. New York: Oxford University Press.

West-Eberhard, M. 2005. Developmental plasticity and the origin of species differences. *Proceedings of the National Academy of Sciences* 102: 6543–6549.

Zhu, R., A. Rebirio, D. Salahub, and S. Kaufmann. 2007. Studying genetic regulatory networks at the molecular level: Delayed reaction stochastic models. *Journal of Theoretical Biology* 246: 725–745.

Chapter 3
Western Atomism and Its Culture-Bound Syndromes

Culture is as much a part of human biology as the enamel on our teeth.
– Robert Boyd.

Summary The stabilization of human cognition via feedback from embedding social and cultural contexts is a dynamic process deeply intertwined with it, constituting, in a sense, the riverbanks directing the flow of a stream of generalized consciousness at different scales: cultural norms and social interaction are synergistic with individual and group cognition and their disorders. A canonical failure mode in atomistic cultures is found to be a "ground state" collapse well-represented by atomistic models of economic interaction that are increasingly characterized as divorced from reality by heterodox economists. That is, high rates of psychopathic and antisocial personality disorder and obsessive-compulsive disorder emerge as culture-bound syndromes particular to Western or Westernizing societies, or to those undergoing social disintegration.

3.1 Introduction

Cognition and its regulation, we will argue, must be viewed as an interacting gestalt, involving not just an atomized individual, but the individual in a rich context that includes embedding social and cultural norms and path-dependent historical trajectory. There can be no cognition without regulation, just as there can be no heartbeat without the control of blood pressure, and no multicellularity without control of rogue cell cancers. Cognitive streams must be contained within regulatory riverbanks.

We implicitly extend the criticisms of Bennett and Hacker (2003) who examined the mereological fallacy of a decontextualization that attributes to "the brain" what is the province of the whole individual. Here, we argue that the "whole individual" incorporates embedding environmental and regulatory settings that, for humans, must include cultural heritage and social interaction.

More specifically, earlier work in this direction (Wallace 2014a,b) explored how information and control theories could be used to examine the stability of cognitive biological processes, in the sense of Maturana and Varela (1980). The failure of such processes is often expressed by the onset of behavioral pathologies and the chronic diseases of aging. Glazebrook and Wallace (2014) apply somewhat similar

R. Wallace, *Computational Psychiatry*, DOI 10.1007/978-3-319-53910-2_3

methods to autism spectrum disorders. Here, we look at certain characteristic mental dysfunctions more specifically in their social and cultural contexts, focusing on observational studies that contrast Western and East Asian populations.

The criteria for antisocial personality disorder (ASPD) in the Diagnostic and Statistical Manual of Mental Disorders (DSM III-R 1987, pp. 344–346)—a standard US medical nosology—lists a set of chronic disruptive, irresponsible, and antisocial behaviors, including lack of remorse and empathy, and similar matters. See Hare et al. (1991) for discussion. Kessler et al. (1994) found, for the USA, a prevalence of about 3.5% of ASPD. Subsequent studies a decade later found similar prevalence (Grant et al. 2004; Compton et al. 2005). By some contrast, Hwu et al. (1989), using DSM III instruments translated into Chinese, found in Taiwan, prevalences of 0.03, 0.07, and 0.14%, respectively, in a large study of village, small town, and Taipei settings. While the "free air of the city" perhaps attracts or enables those with behavioral patterns similar to ASPD, all rates are markedly lower than in the USA.

Similarly, Weissman et al. (1994) used DSM III criteria to quantify obsessive-compulsive disorder (OCD) across a number of international settings. While the lifetime rates for Western societies were found to be approximately 2.3%, in Taiwan, the observed rate was 0.7%. As they put the matter,

> ... [T]he lifetime... and... annual prevalence rates (cases per 100) of obsessive compulsive disorder in seven international communities were remarkably consistent (with the exception of Taiwan)... The prevalence rates for Taiwan were substantially lower than all of the other sites, paralleling Taiwan's low published rates for all psychiatric disorders.

South Korea, in the aftermath of attempts at deculturation carried out during the Japanese occupation between 1910 and 1945, followed by the Korean War, was found to suffer an OCD rate of 1.9% in the study, still lower than the USA at 2.3%.

These observations suggest the necessity of some meditation on the differences between East Asian and Western modes of thinking. Such examination is badly needed: the cultural psychologist Heine (2001) argues that the extreme nature of American individualism suggests that a psychology based on the late twentieth century American research not only stands the risk of developing models that are particular to that culture, but also of developing an understanding of the self that is peculiar in the context of the world's cultures.

In particular, Western atomistic thinking, which pervades a spectrum of disciplines ranging from economics and evolutionary theory to psychology and psychiatry, has deep cultural roots (Wallace 2015, Chap. 1).

Nisbett et al. (2001), following in a long line of research (Markus and Kitayama 1991; Heine 2001), review an extensive literature on empirical studies of basic cognitive differences between individuals raised in East Asian and Western cultural heritages, which they characterize, respectively, as "holistic" and "analytic." They argue:

1. Social organization directs attention to some aspects of the perceptual field at the expense of others.
2. What is attended to influences metaphysics.
3. Metaphysics guides tacit epistemology, that is, beliefs about the nature of the world and causality.

4. Epistemology dictates the development and application of some cognitive processes at the expense of others.
5. Social organization can directly affect the plausibility of metaphysical assumptions, such as whether causality should be regarded as residing in the field vs. in the object.
6. Social organization and social practice can directly influence the development and use of cognitive processes such as dialectical vs. logical ones.

Nisbett et al. (2001) conclude that tools of thought embody a culture's intellectual history, that tools have theories built into them, and that users accept these theories, albeit unknowingly, when they use these tools.

Masuda and Nisbett (2006) find that research on perception and cognition suggests that whereas East Asians view the world holistically, attending to the entire field and relations among objects, Westerners view the world analytically, focusing on the attributes of salient objects. Compared to Americans, East Asians were more sensitive to contextual changes than to focal object changes. These results suggest that there can be cultural variation in what may seem to be basic perceptual processes.

Similarly, Nisbett and Miyamoto (2005) argue that fundamental perceptual processes are influenced by culture. These findings establish a dynamic relationship between the cultural context and perceptual processes. They suggest that perception can no longer be regarded as consisting of processes that are universal across all people at all times.

Wallace (2007) explores analogous dynamics involving inattentional blindness and culture.

A canonical example of how tools of thought embody a culture's intellectual history can be found in mainstream Western economic theory.

3.2 Western Atomistic Economics

The inadequacy of conventional Western economic theorizing has become painfully obvious since the 2008 debacle. Lawson (2010) and Wallace (2015, Chap. 1) provide details and historical context. Important for our purposes is the central point of Lawson's critique of atomistic mathematical models in economics—including, but not limited to, game theory. He identifies a basic mismatch between the sorts of mathematical methods economists employ and the nature of the social, including economic, phenomena that economists seek to illuminate. Their approach usually involves maximization of a simple "utility function" over a distribution of possible strategies. Most fundamentally, economists' methods can be seen to be restricted to closed systems where such maximization can be carried out using standard mathematical techniques. To date, such closures have been found to occur only very rarely in the social realm and we have good reason to suppose they will remain uncommon. In addition, as many heterodox economic theories emphasize, real human dynamics are almost never restricted to atomistic interactions (Lawson 2010; Wallace 2015).

Here, however—and contrarily—we will argue that the symptom spectra of certain psychopathologies may actually characterize such a closed system, one in which the individual (or group) is inherently, or becomes under dynamics of social disintegration, divorced from social or cultural embedding, and engages in a strongly self-referential cognitive dynamic that is most pronounced among atomistic, culturally Western, populations. Different "economic" models may then apply to different characteristic patterns of cognitive dysfunction.

Some development is required.

Cohen (1987) describes control and game theory as follows:

> [H]euristically... the major problem of control theory is to find the maximum of some performance criterion (or criteria), given a set of constraints.... When the objective function is single valued... then one is dealing with optimal control theory. When more than one objective is involved, and the objectives are generally incompatible, then one is dealing with game theory.

The prisoners' dilemma, a classic game theory paradox and a singular example of Lawson's criticisms, is worth special comment. Two prisoners, who engaged jointly in a major crime, have been taken in custody for a relatively minor offense, and are confronted separately and individually by a district attorney. If neither confesses to the major crime, both will serve a short sentence for the minor crime. If both confess, each will receive a longer sentence. If one confesses, and the other does not, the "defector" will be released and the other will serve a very long sentence in prison. Von Neumann game theory predicts—the Nash equilibrium for a one-time game implies—that each will betray the other (Watson 2013).

Field (2014), with a relentless and scathing deconstruction, describes the application of game theory to Cold War nuclear strategy in these terms:

> In the Prisoner's Dilemma played once, for example, the Nash prediction is unambiguous: no cooperation. Defection is the strictly dominant strategy. Experiment... however, provides abundant evidence of positive rates of cooperation. Similar "anomalies" are found in voluntary contribution to public goods games (which are multi person Prisoner's Dilemmas), where one sees positive contribution levels, in the trust game, where one sees positive transfers in both directions, and in many other instances.
>
> Game theory has faced similar predictive failures in its treatment of behavior in the real world. As John von Neumann argued [and, parenthetically, strongly advocated]... its canonical behavioral assumptions predicted devastating conflict between nuclear adversaries... This has not happened...

Field asserts quite forcefully that, because of the disjunction between human behavior and the self-regarding assumptions often used in formal game theory, the latter offers little guidance, normatively or predictively, in thinking about behavior or strategy in a world of potential conflict.

Again, and by contrast, we argue that the behaviorally depauperate dynamics of formal game theory and other closed system mainstream atomistic economic models may actually provide useful characterizations of several Western culture-bound syndromes defined by symptom clusters, including obsessive-compulsive disorder (OCD) and psychopathic and antisocial personality disorders. The latter appear to represent two ends in a spectrum of "natural" and "socially induced" pathology that may generalize to group phenomena in the context of cultural and social collapse.

While the use of game theory to examine psychopathy is far from original (Mokros et al. 2008; Amsel 2007), our route to a more comprehensive formalism is novel. We find game theory and analogous "economic" behaviors can represent forms of relatively simplistic behavioral "ground states" that follow from the developmentally punctuated collapse of essential regulatory systems, both internal and contextual. Ours is, however, most centrally, a "cultural" perspective consistent with the work of Arthur Kleinman and others (Kleinman 1991).

Below, we pursue a minimal line of formal argument for a dauntingly complicated set of nested dynamical structures. Although expressed in only four equations, the embedding mathematical tools are far from elementary.

We begin with an information theory formulation of cognition.

3.3 Cognition as an Information Source

Natural cognitive systems operate at all scales and levels of organization of biological process (Wallace 2012, 2014a). The failure of low level biological cognition in humans is often expressed through early onset of the intractable chronic diseases of senescence (Wallace and Wallace 2010, 2013). Failure of high-order cognition in humans has been the subject of intensive scientific study for over 200 years, with little if any consensus: medical professionals realize that psychological illnesses occur, but they disagree profoundly about their cause and pathology (Johnson-Laird et al. 2006).

This may be something an understatement.

Atmanspacher (2006) argues that theories of high-level cognition are at a point like that of physics 400 years ago, with the basic entities and the relations between them yet to be determined. Further complications arise via the overwhelming influence of culture on both mental process and its dysfunction (e.g., Heine 2001; Kleinman and Cohen 1997).

The stabilization and regulation of high-order cognition for individuals and groups may thus be as complex as such cognition itself.

Some simplification, however, is possible. Cognition can be described in terms of a sophisticated real-time feedback between interior and exterior, necessarily constrained, as Dretske (1994) has noted, by certain asymptotic limit theorems of probability:

> Unless there is a statistically reliable channel of communication between [a source and a receiver]... no signal can carry semantic information... [thus] the channel over which the [semantic] signal arrives [must satisfy] the appropriate statistical constraints of information theory.

The first step in our analysis is a recapitulation of an approach to cognition using the asymptotic limit theorems of information theory (Wallace 2000, 2005a,b, 2007, 2012, 2014a,b).

Atlan and Cohen (1998) argue that the essence of cognition involves comparison of a perceived signal with an internal, learned or inherited picture of the world, and then choice of one response from a much larger repertoire of possible responses. That is, cognitive pattern recognition-and-response proceeds by an algorithmic combination of an incoming external sensory signal with an internal ongoing activity—incorporating the internalized picture of the world—and triggering an appropriate action based on a decision that the pattern of sensory activity requires a response.

Incoming sensory input is thus mixed in an unspecified but systematic manner with internal ongoing activity to create a path of combined signals $x = (a_0, a_1, \ldots, a_n, \ldots)$. Each a_k thus represents some functional composition of the internal and the external. An application of this perspective to a standard neural network is given in Wallace (2005a, p. 34).

This path is fed into a similarly unspecified decision function, h, generating an output $h(x)$ that is an element of one of two disjoint sets B_0 and B_1 of possible system responses. Let

$$B_0 \equiv \{b_0, \ldots, b_k\},$$
$$B_1 \equiv \{b_{k+1}, \ldots, b_m\}.$$

Assume a graded response, supposing that if

$$h(x) \in B_0,$$

the pattern is not recognized, and if

$$h(x) \in B_1,$$

the pattern is recognized, and some action $b_j, k+1 \leq j \leq m$ takes place.

Interest focuses on paths x triggering pattern recognition-and-response: given a fixed initial state a_0, examine all possible subsequent paths x beginning with a_0 and leading to the event $h(x) \in B_1$. Thus $h(a_0, \ldots, a_j) \in B_0$ for all $0 \leq j < m$, but $h(a_0, \ldots, a_m) \in B_1$.

For each positive integer n, take $N(n)$ as the number of high probability paths of length n that begin with some particular a_0 and lead to the condition $h(x) \in B_1$. Call such paths "meaningful," assuming that $N(n)$ will be considerably less than the number of all possible paths of length n leading from a_0 to the condition $h(x) \in B_1$.

The essence of the Shannon–McMillan Theorem that we will invoke below is that the set of low probability paths has, in fact, vanishingly low probability. We will return to this.

Identification of the "alphabet" of the states a_j, B_k may depend on the proper system coarse-graining in the sense of symbolic dynamics (Beck and Schlogl 1995). That is, the larger pattern of behavioral dynamics is projected down onto a simpler, but characteristic, "alphabet" whose combinations and permutations form patterns of "words" and "statements" in a kind of behavioral language "spoken" by the system of interest.

Combining algorithm—i.e., the exact form of the function h—and the details of grammar and syntax are all unspecified in this model. The assumption permitting inference on necessary conditions constrained by the asymptotic limit theorems of information theory is that the finite limit

$$H \equiv \lim_{n \to \infty} \frac{\log[N(n)]}{n} \tag{3.1}$$

both exists and is independent of the path x. Again, $N(n)$ is the number of high probability paths of length n, with low probability paths forming a set of vanishingly small probability.

Call such a pattern recognition-and-response cognitive process ergodic. Not all cognitive processes are likely to be ergodic, implying that H, if it indeed exists at all, is path dependent, although extension to nearly ergodic processes, in a certain sense, seems possible (Wallace 2005a, pp. 31–32).

Invoking the Shannon–McMillan Theorem (Cover and Thomas 2006; Khinchin 1957), we define an adiabatically, piecewise stationary, ergodic information source $\mathbf{X}$ associated with stochastic variates X_j having joint and conditional probabilities $P(a_0, \ldots, a_n)$ and $P(a_n|a_0, \ldots, a_{n-1})$ such that appropriate joint and conditional Shannon uncertainties satisfy the classic relations

$$\begin{aligned} H[\mathbf{X}] &= \lim_{n \to \infty} \frac{\log[N(n)]}{n} \\ &= \lim_{n \to \infty} H(X_n|X_0, \ldots, X_{n-1}) \\ &= \lim_{n \to \infty} \frac{H(X_0, \ldots, X_n)}{n} \end{aligned} \tag{3.2}$$

This information source is defined as dual to the underlying ergodic cognitive process, in the mathematical sense of Wallace (2005a, 2007, 2012).

Again, following Khinchin (1957), the real value of this theorem is that it shows how paths can be divided into just two sets. The first, of high probability, is consistent with inherent patterns of "grammar" and "syntax" (in a large sense) that characterize the information source. The second set, of vanishingly small probability, violates those rules.

"Adiabatic" means that, when the information source is properly parameterized, within continuous "pieces," changes in parameter values take place slowly enough so that the information source remains as close to stationary and ergodic as needed to make the fundamental limit theorems work. "Stationary" means that probabilities do not change in time, and "ergodic" that cross-sectional means converge to long-time averages. Between pieces it is necessary to invoke phase change formalism, a "biological" renormalization that generalizes Wilson's (1971) approach to physical phase transition (Wallace 2005a).

Shannon uncertainties $H(\ldots)$ are cross-sectional law-of-large-numbers sums of the form $-\sum_k P_k \log[P_k]$, where the P_k constitute a probability distribution (Cover and Thomas 2006).

For cognitive systems, an equivalence class algebra can be constructed by choosing different origin points a_0, and defining the equivalence of two states a_m, a_n by the existence of high probability meaningful paths connecting them to the same origin point. Disjoint partition by equivalence class, analogous to orbit equivalence classes for a dynamical system, defines the vertices of a network of cognitive dual languages that interact to actually constitute the system of interest. Each vertex then represents a different information source dual to a cognitive process. This is not a representation of a network of interacting physical systems as such, in the sense of network systems biology (Arrell and Terzic 2010). It is an abstract set of languages dual to the set of cognitive processes of interest, that may become linked into higher order structures.

Topology, however, has long been an object of algebraic study, the so-called algebraic topology, via the fundamental underlying symmetries of geometric spaces. Rotations, mirror transformations, simple ("affine") displacements, and the like uniquely characterize topological spaces, and the networks inherent to cognitive phenomena having dual information sources also have complex underlying symmetries: characterization via equivalence classes defines a groupoid, an extension of the idea of a symmetry group, as summarized by Brown (1987) and Weinstein (1996). Linkages across this set of languages occur via the groupoid generalization of Landau's spontaneous symmetry breaking arguments that will be used below (Landau and Lifshitz 2007; Pettini 2007). See the Mathematical Appendix for a brief summary of basic material on groupoids.

Recognize, however, that we are not constrained in this approach to the Atlan–Cohen model of cognition that, through the comparison with an internal picture of the world, invokes representation. The essential inference is that a broad class of cognitive phenomena—with and without representation—can be associated with a dual information source. That is, cognition inevitably involves choice, choice reduces uncertainty, and reduction of uncertainty implies the existence of an information source whose dynamics are constrained by the asymptotic limit theorems of information theory.

3.4 Environment as an Information Source

Multifactorial cognitive and behavioral systems interact with, affect, and are affected by embedding environments—including culture and social relations—that "remember" interaction by various mechanisms. It is possible to reexpress environmental dynamics in terms of a grammar and syntax that represent the output of an information source—another generalized language.

Some simple examples:

1. The turn-of-the seasons in a temperate climate, for many ecosystems, looks remarkably the same year after year: the ice melts, the migrating birds return, the trees bud, the grass grows, plants and animals reproduce, high summer arrives, the foliage turns, the birds leave, frost, snow, the rivers freeze, and so on.
2. Human interactions take place within fairly well-defined social, cultural, and historical constraints, depending on context: birthday party behaviors are not the same as cocktail party behaviors in a particular social set, but both will be characteristic across different cultural settings.
3. Gene expression during development is highly patterned by embedding environmental context via "norms of reaction" (Wallace and Wallace 2010).

Suppose it possible to coarse-grain the (continuously represented) generalized "ecosystem" at time t, in the sense of symbolic dynamics (e.g., Beck and Schlogl 1995) according to some appropriate finite (or at least countable) partition of the phase space in which each division A_j represent a particular range of numbers of each possible fundamental actor in the generalized ecosystem, along with associated larger system parameters. Of central interest is the set of longitudinal paths—system statements, in a sense—of the form $x(n) = A_0, A_1, \ldots, A_n$ defined in terms of some natural time unit of the system. Thus n corresponds to an again appropriate characteristic time unit T, so that $t = T, 2T, \ldots, nT$.

Let $N(n)$ be the number of possible paths of length n that are consistent with the underlying grammar and syntax of the appropriately coarse-grained embedding ecosystem, in a large sense. As above, the fundamental assumptions are that—for this chosen coarse-graining—$N(n)$, the number of possible grammatical paths, is much smaller than the total number of paths possible, and that, in the limit of (relatively) large n, $H = \lim_{n\to\infty} \log[N(n)]/n$ both exists and is independent of path.

These considerations represent a parallel with parametric statistics. Systems for which the simplifying assumptions are not true will require specialized approaches.

Nonetheless, not all possible ecosystem coarse-grainings—divisions of a continuous system into a finite alphabet-like structure—are likely to work, and different such divisions, even when appropriate, might well lead to different descriptive quasi-languages for the ecosystem of interest. Thus, empirical identification of relevant coarse-grainings for which this theory will work may represent a difficult scientific problem.

Given an appropriately chosen coarse-graining, define joint and conditional probabilities for different ecosystem paths, having the form $P(A_0, A_1, \ldots, A_n)$, $P(A_n|A_0, \ldots, A_{n-1})$, such that appropriate joint and conditional Shannon uncertainties can be defined on them that satisfy Eq. (3.2).

Taking the definitions of Shannon uncertainties as above, and arguing backwards from the latter two parts of Eq. (3.2), it is indeed possible to recover the first, and divide the set of all possible ecosystem temporal paths into two subsets, one very small, containing the grammatically correct, and hence highly probable paths, that we will call "meaningful," and a much larger set of vanishingly low probability.

Learned culture (and its associated patterns of social interaction) contextually constrain possible behaviors, spoken language, body postures, and many other phenotypes. That is, different cultures impose different probability structures on essential matters of living and of the life course trajectory. Even sleep is widely discordant across cultural boundaries. Birth, marriage, death, social conflict, economic exchange, and so on are all strongly patterned by culture, in the context of historical trajectory and social segmentation. Some discussion of these matters in the context of mental disorder can be found in Kleinman and Good (1985), Desjarlais et al. (1995), and the references therein. Boyd and Richerson (2005) provide a more comprehensive introduction.

More generally, as Durham (1991) argues, genes and culture are two distinct but interacting systems of heritage in human populations. Information of both kinds has potential or actual influence over behaviors, creating a real and unambiguous symmetry between genes and phenotypes on the one hand, and culture and phenotypes on the other. Genes and culture are best represented as two parallel tracks of hereditary influence on phenotypes, acting, of course, on markedly different timescales. Human species' identity rests, in no small part, on its unique evolved capabilities for social mediation and cultural transmission, creating, again, high and low probability sets of real-time behavioral sequences.

3.5 Interacting Information Sources

Given a set of information sources that are linked to solve a problem, in the sense of Wilson and Golonka (2013), the "no free lunch" theorem (English 1996; Wolpert and MacReady 1995, 1997) extends a network-based theory (e.g., Arrell and Terzic 2010). Wolpert and Macready show there exists no generally superior computational function optimizer. That is, there is no "free lunch" in the sense that an optimizer pays for superior performance on some functions with inferior performance on others. Gains and losses balance precisely, and all optimizers have identical average performance. In sum, an optimizer has to pay for its superiority on one subset of functions with inferiority on the complementary subset.

This result is known using another description. Shannon (1959) recognized a powerful symmetry between the properties of an information source with a distortion measure and those of a channel. This symmetry is enhanced if we consider channels in which there is a cost associated with the different letters. Solving this problem corresponds to finding a source that is right for the channel and the desired cost. Evaluating the Rate Distortion Function for a source corresponds to finding a channel that is just right for the source and allowed distortion level.

Yet another approach to the same result is through the "Tuning Theorem" of the Mathematical Appendix, which inverts the Shannon Coding Theorem by noting that, formally, one can view the channel as "transmitted" by the signal. Then another kind of "channel capacity" can be defined in terms of the channel probability distribution that maximizes information transmission assuming a fixed message probability distribution.

From the no free lunch argument, Shannon's insight, or the "tuning theorem," it becomes clear that different challenges facing any cognitive system, distributed collection of them, or interacting set of other information sources, that constitute an organism or automaton, must be met by different arrangements of cooperating modules represented as information sources.

It is possible to make a very abstract picture of this phenomenon based on the network of linkages between the information sources dual to the individual "unconscious" cognitive modules (UCM), and those of related information sources with which they interact. That is, a shifting, task-mapped, network of information sources is continually reexpressed: given two distinct problems classes confronting the organism or automaton, there must be two different wirings of the information sources, including those dual to the available UCM, with the network graph edges measured by the amount of information crosstalk between sets of nodes representing the different sources.

Thus "embodied" systems, in the sense of Wilson and Golonka (2013), involve interaction between very general sets of information sources assembled into a "task-specific device" in the sense of Bingham (1988) that is necessarily highly tunable. This mechanism represents a broad evolutionary generalization of the "shifting spotlight" characterizing the global neuronal workspace model of consciousness described in Chap. 1.

3.6 Crosstalk Topologies

A mutual information measure of the inevitable crosstalk between information channels—a kind of energy leakage—is not inherently fixed, but can continuously vary in magnitude. This suggests a parameterized renormalization: the modular network structure linked by crosstalk has a topology depending on the degree of interaction of interest.

Define an interaction parameter ω, a real positive number, and look at geometric structures in the network of interacting cognitive and other information sources defined in terms of linkages set to zero if mutual information is less than, and "renormalized" to unity if greater than, ω. A given ω will define a topologically dependent regime of "giant components" of network elements linked by mutual information greater than or equal to it. See Sect. 1.5 above for details. The emergence of the giant component is well known for random networks.

Now invert the argument: a given topology for the giant component will, in turn, define some critical value, ω_C, so that network elements interacting by mutual information less than that value will be unable to participate, i.e., will be locked out and not be consciously or otherwise perceived. Thus ω is a tunable, syntactically dependent, detection limit that depends critically on the instantaneous topology of the giant component of linked information sources defining the analog to a global broadcast of consciousness.

That topology is the basic tunable syntactic filter across the underlying modular structure, and variation in ω is the only one aspect of more general topological properties that can be described in terms of index theorems, where far more general analytic constraints can become closely linked to the topological structure and dynamics of underlying networks, and, in fact, can stand in place of them (Atiyah and Singer 1963; Hazewinkel 2002). See Sect. 1.6 above for details.

3.7 Punctuated Critical Phenomena

A homology between the information source uncertainty dual to a cognitive process and the free energy density of a physical system arises, in part, from the formal similarity between their definitions in the asymptotic limit. Information source uncertainty can be defined as in the first part of Eq. (3.2). This is quite analogous to the free energy density of a physical system in terms of the thermodynamic limit of infinite volume (Wilson 1971; Wallace 2005a, 2012). Feynman (2000) provides a series of physical examples, based on Bennett's (1988) work, where this homology is an identity, at least for very simple systems. Bennett argues, in terms of idealized irreducibly elementary computing machines, that the information contained in a message can be viewed as the work saved by not needing to recompute what has been transmitted.

It is possible to model a cognitive system interacting with an embedding environment using an extension of the language-of-cognition approach above. Recall that cognitive processes can be formally associated with information sources, and how a formal equivalence class algebra can be constructed for a complicated cognitive system by choosing different origin points in a particular abstract "space" and defining the equivalence of two states by the existence of a high probability meaningful path connecting each of them to some defined origin point within that space.

Recall that disjoint partition by equivalence class is analogous to orbit equivalence relations for dynamical systems, and defines the vertices of a network of cognitive dual languages available to the system: each vertex represents a different information source dual to a cognitive process. The structure creates a large groupoid, with each orbit corresponding to a transitive groupoid whose disjoint union is the full groupoid, and each subgroupoid associated with its own dual information source. Larger groupoids will, in general, have "richer" dual information sources than smaller.

We can now begin to examine the relation between system cognition and the feedback of information from the rapidly changing real-time (as opposed to a slow-time cultural or other) environment, having source uncertainty $\mathcal{H}$.

With each subgroupoid G_i of the (large) cognitive groupoid we can associate a joint information source uncertainty $H(X_{G_i}, Y) \equiv H_{G_i}$, where X is the dual information source of the cognitive phenomenon of interest, and Y that of the embedding human context—largely defined in terms of culture, embedding social structure, and path-dependent historical trajectory. Y is seen as having much slower dynamics that the immediate "environmental" system defining $\mathcal{H}$.

Real-time dynamic responses of a cognitive system can now be represented by high probability paths connecting "initial" multivariate states to "final" configurations, across a great variety of beginning and end points. This creates a similar variety of groupoid classifications and associated dual cognitive processes in which the equivalence of two states is defined by linkages to the same beginning and end states. Thus, we will show, it becomes possible to construct a "groupoid free energy" driven by the quality of rapidly changing, real-time information coming from the embedding ecosystem, represented by the information rate $\mathcal{H}$, taken as a temperature analog.

For humans in particular, $\mathcal{H}$ is a driver for the underlying cognitive processes of interest, here the tunable, shifting, global broadcasts of consciousness as embedded in culture and social relations. The argument-by-abduction from physical theory is, then, that $\mathcal{H}$ constitutes a kind of thermal bath for the processes of culturally channeled cognition. Thus we can, in analogy with the standard approach from statistical physics (Pettini 2007; Landau and Lifshitz 2007), construct a Morse Function by writing a pseudoprobability for the jointly defined information sources X_{G_i}, Y having source uncertainty H_{G_i} as

$$P[H_{G_i}] = \frac{\exp[-H_{G_i}/\kappa\mathcal{H})]}{\sum_j \exp[-H_{G_j}/\kappa\mathcal{H}]} \tag{3.3}$$

where κ is an appropriate dimensionless constant characteristic of the particular system and its linkages to embedding control signals. The sum is over all possible subgroupiods of the largest available symmetry groupoid. Again, compound sources, formed by the (tunable, shifting) union of underlying transitive groupoids, being more complex, will have higher free-energy-density equivalents than those of the base transitive groupoids.

Landau's and Pettini's insights regarding phase transitions in physical systems were that certain critical phenomena take place in the context of a significant alteration in symmetry, with one phase being far more symmetric than the other (Landau and Lifshitz 2007; Pettini 2007). A symmetry is lost in the transition—spontaneous symmetry breaking. The greatest possible set of symmetries in a physical system is that of the Hamiltonian describing its energy states. Usually states accessible at lower temperatures will lack the symmetries available at higher temperatures, so that the lower temperature phase is less symmetric: The randomization of higher temperatures ensures that higher symmetry/energy states will then be accessible to the system. The shift between symmetries is highly punctuated in the temperature index.

A possible Morse Function for invocation of Pettini's topological hypothesis or Landau's spontaneous symmetry breaking is then a "groupoid free energy" F defined by

$$\exp[-F/\kappa\mathcal{H}] \equiv \sum_j \exp[-H_{G_j}/\kappa\mathcal{H}] \tag{3.4}$$

Then, using the free energy-analog F, we apply Landau's spontaneous symmetry breaking arguments, and Pettini's topological hypothesis, to the groupoid associated with the set of dual information sources.

Many other Morse Functions might be constructed here, for example, based on representations of the cognitive groupoid(s). The resulting qualitative pictures would be similar. See the Mathematical Appendix for a summary of results from Morse Theory.

The essential point is that decline in the richness of real-time social and cultural environmental feedback $\mathcal{H}$, or in the ability of that feedback to influence response, as indexed by κ, can lead to punctuated decline in the complexity of cognitive process within the entity of interest, according to this model.

This permits a Landau-analog phase transition analysis in which the quality of incoming information from the embedding ecosystem—feedback—serves to raise or lower the possible richness of an organism's cognitive response to patterns of challenge. If $\kappa\mathcal{H}$ is relatively large—a rich and varied real-time environment, as perceived by the organism—then there are many possible cognitive responses. If, however, noise or simple constraint limit the magnitude of $\kappa\mathcal{H}$, then behavior collapses in a highly punctuated manner to a kind of ground state in which only limited responses are possible, represented by a pathologically simplified cognitive groupoid structure.

3.8 Discussion and Conclusions

We have used a Morse-theoretic extension of results from information theory to explore the dynamics of cognition and its inherently necessary regulation that involves synergistic interpenetration among nested sets of actors, represented here as information sources. These may include dual sources to internal cognitive modules, environmental information, language, culture, social network, socioeconomic affordances and limitations, and so on.

Two factors determine the possible range of real-time cognitive response in the model. These are the magnitude of the environmental feedback control signal and the inherent structural richness of the underlying cognitive groupoid. If that richness is lacking—if the available topologies of internal ω-driven connections is limited—then even very high levels of $\kappa\mathcal{H}$ may not be adequate to activate appropriate behavioral responses to important regulatory feedback signals. Glazebrook and Wallace (2014), in fact, examine autism spectrum disorders from this viewpoint.

Here, we are particularly interested in a "ground state collapse" in which cultural and social restraints fail to stabilize individual (and perhaps group) behaviors. A key observation, we contend, is the marked difference in observed prevalence of obsessive-compulsive disorder and antisocial personality disorder(s) among "East Asian" Taiwanese and "Western" populations.

Using the theoretical framework above, psychopathic behaviors would be viewed as intrinsic to culturally modulated brain development, represented by failures in the topological ω-mechanisms. Acquired antisocial personality disorder would then stem from imposed psychosocial stresses during the life course, weakening $\kappa\mathcal{H}$ in the phase transition model.

The resulting ground state pathologies seem described by surprisingly uncomplicated behaviors arising from simplistic self-interest calculations that apparently follow something much like von Neumann's game theory model of a Cold War preemptive thermonuclear exchange, or the calculations of a mainstream economist who analyzes criminal behavior (Becker 1968; Wallace and Fullilove 2014). Amsel's (2007) work indicates that OCD behaviors may involve analogous, but more subtle, regulatory collapse leading again to simplistic closed-system "cost" analysis. High rates of such pathologies are to be viewed, taking Kleinman's (1991) perspective, as culture-bound syndromes peculiar to, or at least significantly enhanced in, atomistic Western or Westernizing societies, or, in the case of Korea, communities exposed to severe colonial depredation and war.

The underlying cultural dynamics have been the subject of much past commentary. Maxim Gorky (1972), in his classic *City of the Yellow Devil* describing New York City, writes:

> The youth leaning against the lamppost shakes his head from time to time. His hungry teeth are tightly clenched. I believe I understand what he is thinking of, what he wants... to possess enormous hands of frightful strength and wings on his back, that is what he wants, I believe. So that, soaring 1 day over the city, he may reach down with hands like steel levers and reduce the whole to a heap of rubbish and ashes, mixing bricks and pearls, gold and the flesh of slaves, glass and millionaires, dirt, idiots, temples, the dirt-poisoned trees, and these foolish multi-storeyed skyscrapers, everything, the whole city into one heap, into a dough compounded of dirt and the blood of people—into a loathsome chaos. This frightful wish is as natural in this youth's brain as a sore on the body of a sick man. Where there is much work for slaves, there can be no place for free, creative thought, and only the ideas of destruction, the poisonous flowers of vengeance, the turbulent protest of the brute beast can flourish. This is understandable—if you warp a man's soul you must not expect mercy from him.

The French-trained psychiatrist Frantz Fanon (1966), describing the impact of Western colonialism, characterized the underlying mechanism as follows:

> The colonized man will first manifest this aggressiveness which has been deposited in his bones against his own people. This is the period when the niggers beat each other up, and the police and magistrates do not know which way to turn when faced with the astonishing waves of crime...

Effective untangling of such developmental knots will require more than interventions at the individual level. Necessary as they may be, they are never sufficient.

In closing, it could be conjectured that the perspective expressed in the chapter title, "Western atomism and its culture-bound syndromes," might generalize across cultures, perhaps in the direction of "East Asian collectivism and its culture-bound syndromes," a matter for further study.

Acknowledgements The author thanks Dr. D.N. Wallace for fruitful discussions and two reviewers for comments useful in revision.

References

Amsel, L. 2007. Towards a mathematical psychiatry: Modeling repetition compulsion with game theory. In *11th Annual Meeting*. Association for the Scientific Study of Consciousness.

Arrell, D., and A. Terzic. 2010. Network systems biology for drug discovery. *Clinical Pharmacology and Therapeutics* 88: 120–125.

Atiyah, M., and I. Singer. 1963. The index of elliptical operators on compact manifolds. *Bulletin of the American Mathematical Society* 69: 322–433.

Atlan, H., and I. Cohen. 1998. Immune information, self-organization, and meaning. *International Immunology* 10: 711–717.

Atmanspacher, H. 2006. Toward an information theoretical implementation of contextual conditions for consciousness. *Acta Biotheoretica* 54: 157–160.

Beck, C., and F. Schlogl. 1995. *Thermodynamics of Chaotic Systems*. New York: Cambridge University Press.

Becker, C. 1968. Crime and punishment: An economic approach. *Journal of Political Economy* 76: 169–217.

Bennett, C. 1988. Logical depth and physical complexity. In *The Universal Turing Machine: A Half-Century Survey*, ed. R. Herkin, 227–257. New York: Oxford University Press.

Bennett, M., and P. Hacker. 2003. *Philosophical Foundations of Neuroscience*. London: Blackwell.

Bingham, G. 1988. Task-specific devices and the perceptual bottleneck. *Human Movement Science* 7: 225–264.

Boyd, R., and P. Richerson. 2005. *The Origin and Evolution of Cultures*. New York: Oxford University Press.

Brown, R. 1987. From groups to groupoids: a brief survey. *Bulletin of the London Mathematical Society* 19: 113–134.

Cohen, Y. 1987. *Preface, Applications of Control Theory in Ecology*. New York: Springer.

Compton, W., et al. 2005. Prevalence, correlates, and comorbidity of DSM-IV antisocial personality syndromes and alcohol and specific drug disorders in the United States: Results from the national epidemiologic survey on alcohol and related conditions. *Journal of Clinical Psychiatry* 66: 677–685.

Cover, T., and J. Thomas. 2006. *Elements of Information Theory*, 2nd ed. New York: Wiley.

Desjarlais, R., et al. 1995. *World Mental Health*. New York: Oxford University Press.

Dretske, F. 1994. The explanatory role of information. *Philosophical Transactions of the Royal Society A* 349: 59–70.

DSM III. 1987. *Diagnostic and Statistical Manual of Mental Disorders*. Washington, DC: American Psychiatric Association.

Durham, W. 1991. *Coevolution: Genes, Culture, and Human Diversity*. Palo Alto, CA: Stanford University Press.

English, T. 1996. Evaluation of evolutionary and genetic optimizers: No free lunch. In *Evolutionary Programming V: Proceedings of the Fifth Annual Conference on Evolutionary Programming*, ed. L. Fogel, P. Angeline, and T. Back, 163–169. Cambridge, MA: MIT.

Fanon, F. 1966. *The Wretched of the Earth*. Boston, MA: Beacon.

Feynman, R. 2000. *Lectures on Computation*. Boulder, CO: Westview.
Field, A. 2014. Schelling, von Neumann, and the event that didn't occur. *Games* 5: 53–89.
Glazebrook, J.F., and R. Wallace. 2014. Pathologies in functional connectivity, feedback control and robustness: A global workspace perspective on autism spectrum disorders. *Cognitive Processing*. doi: 10.1007/s10339-014-0636-y.
Gorky, M. 1972[1906]. *The City of the Yellow Devil*. Moscow USSR: Progress Publishers.
Grant, B., et al. 2004. Prevalence, correlates, and disability of personality disorders in the United States: Results from the national epidemiologic survey on alcohol and related conditions. *Journal of Clinical Psychiatry* 7: 948–958.
Hare, R., et al. 1991. Psychopathy and the DSM-IV Criteria for antisocial personality disorder. *Journal of Abnormal Psychology* 100: 391–398.
Hazewinkel, M. 2002. *Index Formulas, Encyclopedia of Mathematics*. New York: Springer.
Heine, S. 2001. Self as cultural product: An examination of East Asian and North American selves. *Journal of Personality* 69: 881–906.
Hwu, H., et al. 1989. Prevalence of psychiatric disorders in Taiwan defined by the Chinese diagnostic interview schedule. *Acta Psychiatrica Scandinavia* 79: 136–147.
Johnson-Laird, P., F. Mancini, and A. Gangemi. 2006. A hyperemotional theory of psychological illness. *Psychological Review* 113: 822–841.
Kessler, R., et al. 1994. Lifetime and 12-month prevalence of DSM-III-R psychiatric disorders in the United States. Results from the National Comorbidity Survey. *Archives of General Psychiatry* 51: 8–19.
Khinchin, A. 1957. *The Mathematical Foundations of Information Theory*. New York: Dover.
Kleinman, A. 1991. *Rethinking Psychiatry: From Cultural Category to Personal Experience*. New York: Free Press.
Kleinman, A., and A. Cohen. 1997. Psychiatry's global challenge. *Scientific American* 276(3): 86–89.
Kleinman, A., and B. Good. 1985. *Culture and Depression*. San Francisco: California University Press.
Landau, L., and E. Lifshitz. 2007. *Statistical Physics*, 3rd ed., Part I. New York: Elsevier.
Lawson, T. 2010. Really reorienting modern economics. In *INET Conference*, London, April 8–11.
Markus, H., and S. Kitayama. 1991. Culture and the self-implications for cognition, emotion, and motivation. *Psychological Review* 98: 224–253.
Masuda, T., and R. Nisbett. 2006. Culture and change blindness. *Cognitive Science* 30: 381–399.
Maturana, H., and F. Varela. 1980. *Autopoiesis and Cognition*. Dordrecht: Reidel.
Mokros, A., et al. 2008. Diminished cooperativeness of psychopaths in a Prisoner's Dilemma Game yields higher rewards. *Journal of Abnormal Psychology* 117: 406–413.
Nisbett, R., and Y. Miyamoto. 2005. The influence of culture: Holistic versus analytic perception. *Trends in Cognitive Science* 9: 467–473.
Nisbett, R., et al. 2001. Culture and systems of thought: Holistic vs. analytic cognition. *Psychological Review* 108: 291–310.
Pettini, M. 2007. *Geometry and Topology in Hamiltonian Dynamics and Statistical Mechanics*. New York: Reidel.
Shannon, C. 1959. Coding theorems for a discrete source with a fidelity criterion. *Institute of Radio Engineers International Convention Record* 7: 142–163.
Wallace, R. 2000. Language and coherent neural amplification in hierarchical systems: Renormalization and the and the dual information source of a generalized spatiotemporal stochastic resonance. *International Journal of Bifurcation and Chaos* 10: 493–502.
Wallace, R. 2005a. *Consciousness: A Mathematical Treatment of the Global Neuronal Workspace Model*. New York: Springer.
Wallace, R. 2005b. A global workspace perspective on mental disorders. *Theoretical Biology and Medical Modelling* 2: 49.
Wallace, R. 2007. Culture and inattentional blindness: A global workspace perspective. *Journal of Theoretical Biology* 245: 378–390.

Wallace, R. 2012. Consciousness, crosstalk, and the mereological fallacy: An evolutionary perspective. *Physics of Life Reviews* 9: 426–453.
Wallace, R. 2014a. Cognition and biology: Perspectives from information theory. *Cognitive Processing* 15: 1–12.
Wallace, R. 2014b. Metabolic free energy and biological codes: A 'Data Rate Theorem' aging model. *Bulletin of Mathematical Biology*. doi:10.1007/s11538-014-0013-0.
Wallace, R. 2015. *An Ecosystem Approach to Economic Stabilization: Escaping the Neoliberal Wilderness*, vol. 21. Routledge Advances in Heterodox Economics. London: Routledge.
Wallace, R., and R. Fullilove. 2014. State policy and the political economy of criminal enterprise: Mass incarceration and persistent organized hyperviolence in the USA. *Structural Change and Economic Dynamics* 31: 17–35.
Wallace, R., and D. Wallace. 2010. *Gene Expression and its Discontents: The Social Production of Chronic Disease*. New York: Springer.
Wallace, R., and D. Wallace. 2013. *A Mathematical Approach to Multilevel, Multiscale Health Interventions: Pharmaceutical Industry Decline and Policy Response*. London: Imperial College Press.
Watson, J. 2013. *Strategy: An Introduction to Game Theory*. New York: W. W. Norton.
Weissman, M., et al. 1994. The cross national epidemiology of obsessive compulsive disorder. The cross national collaborative group. *Journal of Clinical Psychiatry* 55(Suppl.): 5–10.
Weinstein, A. 1996. Groupoids: unifying internal and external symmetry. *Notices of the American Mathematical Association* 43: 744–752.
Wilson, K. 1971. Renormalization group and critical phenomena I. Renormalization group and the Kadanoff scaling picture. *Physical Review B* 4: 3174–3183.
Wilson, A., and S. Golonka. 2013. Embodied cognition is not what you think it is. *Frontiers in Psychology*. doi: 10.3389/fpsyg.2013.00058.
Wolpert, D., and W. MacReady. 1995. No free lunch theorems for search. Santa Fe Institute, SFI-TR-02-010.
Wolpert, D., and W. MacReady. 1997. No free lunch theorems for optimization. *IEEE Transactions on Evolutionary Computation* 1: 67–82.

Chapter 4
Environmental Induction of Neurodevelopmental Disorders

Summary Chemical exposures, pre- and neonatal infections, psychosocial stress, and the cross-generational cultural and epigenetic impacts of these and other toxicants become an integrated, sometimes synergistic, signal that can overwhelm essential neurodevelopmental regulation. We characterize that dynamic through statistical models based on the asymptotic limit theorems of control and information theories. Schizophrenia and autism emerge as two different "phases" of pathological neurodevelopmental "condensations" that impair the dynamic, shifting global workspace of normal mental function.

> Early neglect and life course environmental insults... can lead to impaired neuronal responsiveness and symptoms of profound prefrontal cortical dysfunction, providing a direct link between the environment and the cognitive impairments observed in psychotic syndrome. (van Os et al. 2010)

4.1 Introduction

Growing horror over the possible relation between the rapid spread of mosquito-borne Zika virus and increased rates of microcephaly among children born to women infected during pregnancy (Hayden 2016; Garcez et al. 2016) sharply focuses attention on how environmental exposures, in a large sense, might trigger neurodevelopmental disorders. Here, we extend the perspective of Wallace (2015a), which focused on the pathologies of aging, to examine the induction of developmental disorders in the presence of environmental disruption, in a large sense.

The underlying importance of neurodevelopment in the etiology of serious mental disorder has been described cogently by Corbin et al. (2008) who conclude that

> ...[U]nraveling the mechanisms of neural progenitor cell diversity in the brain has tremendous clinical importance... [D]efects in any of these processes can have devastating and long lasting consequences on brain function...
>
> ...[A]bnormal development of interneurons may be an underlying causative factor, or contribute to the phenotype of a variety of developmental disorders... [including] autism spectrum disorders... [and] schizophrenia...

R. Wallace, *Computational Psychiatry*, DOI 10.1007/978-3-319-53910-2_4

Similarly, Tiberi et al. (2012) describe how the cerebral cortex is composed of hundreds of different types of neurons, which underlie its ability to perform highly complex neural processes. How cortical neurons are generated during development constitutes a major challenge in developmental neurosciences with important implications for brain repair and diseases. Cortical neurogenesis is dependent on intrinsic and extrinsic clues, which interplay to generate cortical neurons at the right number, time, and place. Recent evidence, in their view, indicates that most classical morphogens, produced by various neural and non-neural sources throughout embryonic development, contribute to the master control and fine tuning of cortical neurogenesis. They conclude that the molecular control of cortical neurogenesis involves the interplay of intrinsic and extrinsic cues that coordinate the pattern of neural progenitor division and differentiation.

Rapoport et al. (2012) find the neurodevelopmental model positing illness as the end stage of abnormal neurodevelopmental processes that began years before the onset of the illness. Environmental risk factors such as urbanicity, childhood trauma, and social adversity have received strong replication with marked phenotypic nonspecificity pointing to common brain development pathways across disorders. The neurodevelopmental model of schizophrenia has long existed as a model for other childhood-onset conditions, including attention deficit hyperactivity disorder, intellectual deficiency, autism spectrum disorders, and epilepsy.

Rapoport et al. specifically identify infection/famine, placental pathology, low birth weight, urban environment, childhood trauma, and ethnic minority/immigrant status in disease etiology, concluding that in the central nervous system, neuronal proliferation, cell migration, morphological and biochemical differentiation, and circuit formation all depend on cell and cell–environment interactions that control developmental process, and so can cause altered trajectories. Their Fig. 1 provides a summary schematic.

A parallel line of argument explores mitochondrial abnormalities that are closely associated with both schizophreniform and autism spectrum disorders.

Ben-Shachar (2002) finds mitochondrial impairment could provide an explanation for the broad spectrum of clinical and pathological manifestations in schizophrenia. Several independent lines of evidence, Ben-Shachar asserts, suggest an involvement of mitochondrial dysfunction in the disorder, including altered cerebral energy metabolism, mitochondrial polyplasia, dysfunction of the oxidative phosphorylation system, and altered mitochondrial related gene expression. They conclude that the interaction between dopamine, a predominant etiological factor in schizophrenia, and mitochondrial respiration is a possible mechanism underlying the hyper- and hypo-activity cycling in schizophrenia.

Prabakaran et al. (2004) claim that almost half the altered proteins identified by a brain tissue proteomics analysis of samples from schizophrenic patients were associated with mitochondrial function and oxidative stress responses. They propose that oxidative stress and the ensuing cellular adaptations are linked to the schizophrenia disease process.

Shao et al. (2008) similarly find evidence of mitochondrial dysfunction in schizophrenia. Likewise, Scaglia (2010) suggests involvement of mitochondrial dysfunction in schizophrenia and argues that mechanisms of dysfunctional cellular energy metabolism underlie the pathophysiology of major subsets of psychiatric disorders.

Clay et al. (2011) point to an underlying dysfunction of mitochondria in bipolar disorder and schizophrenia including (1) decreased mitochondrial respiration; (2) changes in mitochondrial morphology; (3) increases in mitochondrial DNA (mtDNA) polymorphisms and in levels of mtDNA mutations; (4) downregulation of nuclear mRNA molecules and proteins involved in mitochondrial respiration; (5) decreased high-energy phosphates and decreased pH in the brain; and (6) psychotic and affective symptoms, and cognitive decline in mitochondrial disorders. They conclude that understanding the role of mitochondria, both developmentally and in the ailing brain, is of critical importance to elucidate pathophysiological mechanisms in psychiatric disorders.

There is likewise considerable and growing evidence for mitochondrial mechanisms in autism spectrum disorders (ASD).

Palmieri and Persico (2010) find ASD often associated with clinical, biochemical, or neuropathological evidence of altered mitochondrial function. The majority of autistic patients display functional abnormalities in mitochondrial metabolism seemingly secondary to pathophysiological triggers. Thus, in their view, mitochondrial function may play a critical role not just in rarely causing the disease, but also in frequently determining to what extent different prenatal triggers will derange neurodevelopment and yield abnormal postnatal behavior.

Giulivi et al. (2010) similarly assert that impaired mitochondrial function may influence processes highly dependent on energy, such as neurodevelopment, and contribute to autism. In their study, children with autism were more likely to have mitochondrial dysfunction, mtDNA overreplication, and mtDNA deletions than typically developing children.

A long series of studies by Rossignol and Frye (2010) find evidence accumulating that ASD is characterized by certain physiological abnormalities, including oxidative stress, mitochondrial dysfunction, and immune dysregulation/inflammation. Recent studies, they conclude, have reported these abnormalities in brain tissue derived from individuals diagnosed with ASD as compared to brain tissue derived from control individuals, suggesting that ASD has a clear biological basis with features of known medical disorders.

Goh et al. (2014) argue that impaired mitochondrial function impacts many biological processes that depend heavily on energy and metabolism and can lead to a wide range of neurodevelopmental disorders, including ASD. Although, in their view, evidence that mitochondrial dysfunction is a biological subtype of ASD has grown in recent years, no study had previously demonstrated evidence of mitochondrial dysfunction in brain tissue in vivo in a large, well-defined sample of individuals with ASD. Their use of sensitive imaging technologies allowed them to identify in vivo a biological subtype of ASD with mitochondrial dysfunction. Lactate-positive voxels in their sample were detected most frequently in the cingulate gyrus, a structure that supports higher order control of thought, emotion, and behavior, and one in which both anatomical and functional disturbances have been reported previously in ASD.

For neurodevelopment, control of gene expression is everything, and mechanisms by which environmental factors interfere with control are of essential clinical and epidemiological concern.

Here, we will describe statistical models of developmental failure based on the asymptotic limit theorems of control and information theories that may provide new tools in exploring such mechanisms. The models are analogous to more familiar empirical least-squares regression and may permit deep scientific inference arising from comparison of similar systems under different, or different systems under similar, experimental or observational circumstances.

4.2 A Control Theory Model

It is well understood that there is no gene expression without regulation. This implies that gene expression is inherently unstable in the formal control theory sense of the Data Rate Theorem (Nair et al. 2007), and must be stabilized by provision of control information at a critical rate. Failure to provide control information at or above that rate initiates characteristic modes of system failure that, for neural systems, are expressed as developmental disorders. More explicitly, assuming an approximate nonequilibrium steady state, the simplest "regression" model of deviations from that state—described in terms of an n-dimensional vector of observables x_t at time t—has the form

$$x_{t+1} = \mathbf{A}x_t + \mathbf{B}u_t + W_t \tag{4.1}$$

where x_{t+1} is the state at time $t + 1$, u_t is the imposed n-dimensional control signal vector at time t, W_t is an added noise signal, and $\mathbf{A}$ and $\mathbf{B}$ are, in this approximation, fixed $n \times n$ matrices. See Fig. 4.1 for a schematic.

The Data Rate Theorem (Nair et al. 2007) states that, for an inherently unstable system, the control information represented by the vector u_t must be provided at a rate $\mathcal{H}$ that is greater than the rate at which the system produces "topological information." For the system of Eq. (4.1) and Fig. 4.1, that rate is given as

$$\mathcal{H} > \log[|\det(\mathbf{A}^u)|] \equiv \alpha_0 \tag{4.2}$$

where det is the determinant and $\mathbf{A}^u$ is the component submatrix of $\mathbf{A}$ that has eigenvalues ≥ 1.

An alternate derivation of Eq. (4.2) is given in the Mathematical Appendix.

Generalization to more complex inherently unstable systems in the context of a scalar integrated environmental insult ρ—for example, taken as the magnitude of the largest vector of an empirical principal component analysis—suggests that Eq. (4.2) will become something like

$$\mathcal{H}(\rho) > f(\rho)\alpha_0 \tag{4.3}$$

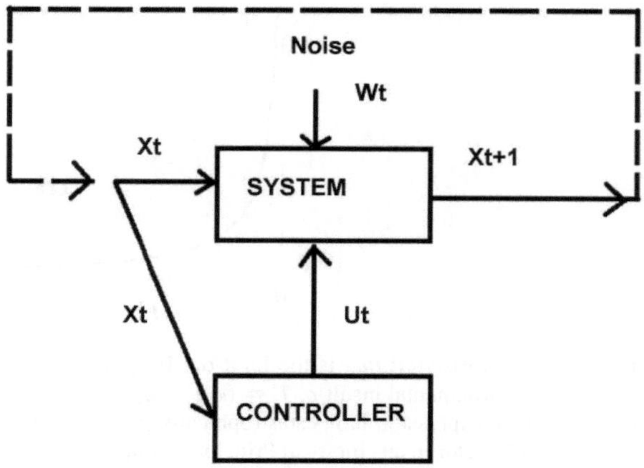

Fig. 4.1 "Regression model" for a control system near a nonequilibrium steady state. x_t is system output at time t, u_t the control signal, and W_t an added noise term

$f(0)\alpha_0$ is then interpreted as the rate at which the system generates topological information in the absence of an integrated environmental exposure.

What are the forms of $\mathcal{H}(\rho)$ and $f(\rho)$? In the Mathematical Appendix we calculate $\mathcal{H}(\rho)$ as the "cost" of control information, given the "investment" ρ, using a classic Black–Scholes approximation (Black and Scholes 1973). To first order,

$$\mathcal{H}(\rho) \approx \kappa_1 \rho + \kappa_2 \tag{4.4}$$

where the κ_i are positive or zero.

If we take the same level of approximation, $f(\rho)$ in Eq. (4.3) can be similarly expressed as $\kappa_3 \rho + \kappa_4$ so that the stability condition is

$$\mathcal{T} \equiv \frac{\kappa_1 \rho + \kappa_2}{\kappa_3 \rho + \kappa_4} > \alpha_0 \tag{4.5}$$

For small ρ the stability requirement is $\kappa_2/\kappa_4 > \alpha_0$, and at high ρ it becomes $\kappa_1/\kappa_3 > \alpha_0$. If $\kappa_2/\kappa_4 \gg \kappa_1/\kappa_3$, then at some intermediate value of ρ the essential inequality may be violated, leading to failure of neurodevelopmental regulation. See Fig. 4.2.

Fetal, child, and indeed adult, developmental trajectories are embedded not only in environments of direct exposure, but also of multimodal inheritance, both through cross-generational gene methylation and other biochemical mechanisms and via sociocultural influences. It is, however, implicit that direct environmental exposures, inherited gene methylation, sociocultural inheritance, and other important factors must interact along the developmental trajectory. Thus, rather than a simple scalar we are confronted by an $m \times m$ matrix having elements $\rho_{i,j}$ $i, j = 1 \ldots m$.

Square matrices of order m, however, have m scalar invariants, m real numbers that characterize the matrix regardless of how it is expressed in different coordinate systems. The first is the trace, and the last $\pm$ the determinant. In general, the

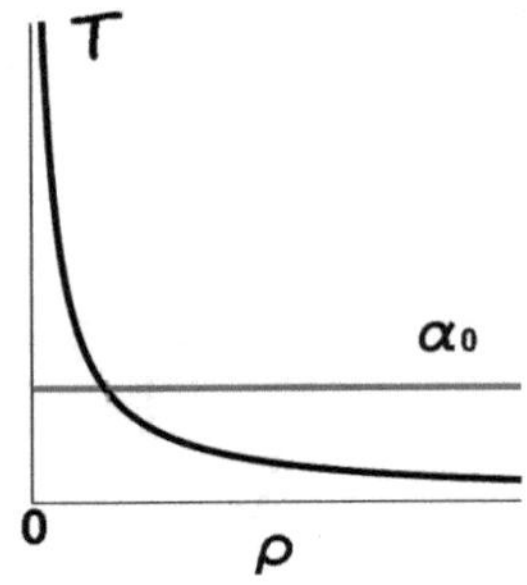

Fig. 4.2 The *horizontal line* is the limit α_0. If $\kappa_2/\kappa_4 \gg \kappa_1/\kappa_3$, at some intermediate value of integrated environmental insult ρ, $\mathcal{T} = (\kappa_1\rho + \kappa_2)/(\kappa_3\rho + \kappa_4)$ falls below criticality, and control of neural gene expression fails catastrophically. ρ itself might be calculated as the magnitude of the "volume" vector in an empirical Principal Component Analysis or through a more complicated model that explicitly accounts for different epigenetic inheritances and their cross-influences

invariants are the coefficients of the characteristic polynomial $\mathcal{P}(\lambda)$:

$$\mathcal{P}(\lambda) = \det(\rho - \lambda I) = \lambda^m + r_1\lambda^{m-1} + \cdots + r_{m-1}\lambda + r_m \tag{4.6}$$

where λ is a parameter that is an element of some ring, det is the determinant, and I the $m \times m$ identity matrix. Note that λ may, in fact, be taken as the matrix ρ itself, since square matrices form a ring, in which case the relation is a matrix polynomial $\mathcal{P}(\rho) = 0 \times I$.

For an $m \times m$ matrix we have invariants $r_1, \ldots, r_m$ and an appropriate scalar "ρ" in Eq. (4.6)—determining the "temperature" $\mathcal{T}$—is then a monotonic increasing function of the r_i:

$$\hat{\rho} = \hat{\rho}(r_1, \ldots, r_m) \tag{4.7}$$

so that

$$\mathcal{T}(\hat{\rho}) = \frac{\kappa_1\hat{\rho} + \kappa_2}{\kappa_3\hat{\rho} + \kappa_4} \tag{4.8}$$

We have invoked the "Rate Distortion Manifold" of Glazebrook and Wallace (2009)—formulated as a "generalized retina" in Wallace and Wallace (2010)—to project a complicated "information manifold" down onto a lower dimensional "tangent space" tuned across that manifold in such a way as to preserve most of the underlying information. Here, we assume a scalar tangent space. Higher dimensional structures are possible in a standard manner at the cost of some considerable increase in mathematical overhead.

What are the dynamics of $\mathcal{T}(\hat{\rho})$ under stochastic circumstances? We explore this by examining how a control signal u_t in Fig. 4.1 *is expressed in the system response* x_{t+1}. More explicitly, we suppose it possible to deterministically retranslate a sequence of system outputs $X^i = x_1^i, x_2^i, \ldots$ into a sequence of possible control signals $\hat{U}^i = \hat{u}_0^i, \hat{u}_1^i, \ldots$ and then compare that sequence with the original control sequence $U^i = u_0^i, u_1^i, \ldots$. The difference between them is a real number measured by a chosen distortion measure, enabling definition of an average distortion

$$< d >= \sum_i p(U^i)d(U^i, \hat{U}^i) \tag{4.9}$$

where (1), $p(U^i)$ is the probability of the sequence U^i (2), $d(U^i, \hat{U}^i)$ is the distortion between U^i and $\hat{U}^i$ and (3), the sequence of control signals has been deterministically reconstructed from the system output.

It then becomes possible to apply a classic Rate Distortion Theorem (RDT) argument. According to the RDT, there exists a Rate Distortion Function (RDF) that determines the minimum channel capacity, $R(D)$, necessary to keep the average distortion $< d >$ below some fixed limit D (Cover and Thomas 2006). Based on Feynman's (2000) interpretation of information as a form of (free) energy, we can then construct a Boltzmann-like pseudoprobability density in the "temperature" $\mathcal{T}$ as

$$dP(R, \mathcal{T}) = \frac{\exp[-R/\mathcal{T}]dR}{\int_0^\infty \exp[-R/\mathcal{T}]dR} \tag{4.10}$$

since higher $\mathcal{T}$ necessarily implies greater channel capacity.

The integral in the denominator is essentially a statistical mechanical partition function, and we can then define a "free energy" Morse Function F (Pettini 2007) as

$$\exp[-F/\mathcal{T}] = \int_0^\infty \exp[-R/\mathcal{T}]dR = \mathcal{T} \tag{4.11}$$

so that $F(\mathcal{T}) = -\mathcal{T}\log[\mathcal{T}]$.

Then an entropy-analog can also be defined as the Legendre transform of F:

$$\mathcal{S} \equiv F(\mathcal{T}) - \mathcal{T}dF/d\mathcal{T} = \mathcal{T} \tag{4.12}$$

As a first approximation, Onsager's treatment of nonequilibrium thermodynamics (de Groot and Mazur 1984) can be applied, so that system dynamics are driven by the gradient of $\mathcal{S}$ in essential parameters—here $\mathcal{T}$—under conditions of noise. This gives a stochastic differential equation

$$d\mathcal{T}_t \approx (\mu d\mathcal{S}/d\mathcal{T})dt + \beta\mathcal{T}_t dW_t = \mu dt + \beta\mathcal{T}_t dW_t \tag{4.13}$$

where μ is a "diffusion coefficient" representing the efforts of the underlying control mechanism, and β is the magnitude of an inherent impinging white noise dW_t in the context of volatility, i.e., noise proportional to signal.

Somewhat heuristically, applying the Ito chain rule to $\log(\mathcal{T})$ in Eq. (4.13), via Jensen's inequality for a concave function, a nonequilibrium steady state (nss) expectation for $\mathcal{T}$ can be calculated as

$$E(\mathcal{T}_t) \geq \frac{\mu}{\beta^2/2} \tag{4.14}$$

Again, μ is interpreted as indexing the attempt by the embedding control apparatus to impose stability—raise $\mathcal{T}$. Thus impinging noise can significantly increase the probability that $\mathcal{T}$ falls below the critical limit of Fig. 4.2, initiating a control failure.

However, $E(\mathcal{T})$ is an expectation, so that, in this model, there is always some nonzero probability that $\mathcal{T}$ will fall below the critical value α_0 in the multimodal expression for $\mathcal{T}(\hat{\rho})$: sporadic control dysfunctions have not been eliminated. Raising μ and lowering β decreases their probability, but will not drive it to zero in this model, a matter of some importance for population-rates of neurodevelopmental disorders.

4.3 A "Cognitive" Model

A different approach to the dynamics of neurodevelopmental regulation applies the "cognitive paradigm" of Atlan and Cohen (1998), who recognized that the immune response is not merely an automatic reflex, but involves active choice of a particular response to insult from a larger repertoire of possible responses. Choice reduces uncertainty and implies the existence of an underlying information source as argued in earlier chapters.

Given an information source associated with an inherently unstable, rapidly acting cognitive neurodevelopmental control system—called "dual" to it—an equivalence class algebra can be constructed by choosing different system origin states a_0 and defining the equivalence of two subsequent states at times $m, n > 0$, written as a_m, a_n, by the existence of high probability meaningful paths connecting them to the same origin point. Disjoint partition by equivalence class, analogous to orbit equivalence classes in dynamical systems, defines a symmetry groupoid associated with the cognitive process. Groupoids are deep generalizations of the group concept in which there is not necessarily a product defined for each possible element pair (Weinstein 1996).

The equivalence classes define a set of cognitive dual information sources available to the inherently unstable neurodevelopment regulation system, creating a large groupoid, with each orbit corresponding to a transitive groupoid whose disjoint union is the full groupoid. Each subgroupoid is associated with its own dual information source, and larger groupoids will have richer dual information sources than smaller.

Let X_{G_i} be the control system's dual information source associated with the groupoid element G_i, and let Y be the information source associated with embedding "normal" environmental variation that impinges on development. Wallace (2012, 2015b) gives details of how environmental regularities imply the existence of an environmental information source that, for humans, particularly includes cultural and socioeconomic factors (e.g., Wallace 2015c).

We can again construct a "free energy" Morse Function (Pettini 2007). Let $H(X_{G_i}, Y) \equiv H_{G_i}$ be the joint uncertainty of the two information sources. Another Boltzmann-like pseudoprobability can then be written as

$$P[H_{G_i}] = \frac{\exp[-H_{G_i}/\mathcal{T}]}{\sum_j \exp[-H_{G_j}/\mathcal{T}]} \tag{4.15}$$

$\mathcal{T}$ is the "temperature" from Eq. (4.9), via the $\hat{\rho}$ of Eq. (4.7), and the sum is over the different possible cognitive modes of the full system.

A new Morse Function $\mathcal{F}$ is defined by

$$\exp[-\mathcal{F}/\mathcal{T}] \equiv \sum_j \exp[-H_{G_j}/\mathcal{T}] \tag{4.16}$$

Given the inherent groupoid structure as a generalization of the simple symmetry group, it becomes possible to apply an extension of Landau's picture of phase transition (Pettini 2007). In Landau's "spontaneous symmetry breaking", phase transitions driven by temperature changes occur as alteration of system symmetry, with higher energies at higher temperatures being more symmetric.

For this model, the shift between symmetries is highly punctuated in the "temperature" index $\mathcal{T}$ under the Data Rate Theorem for unstable control systems. Typically, there are only a very limited number of possible phases, which may or may not coexist under particular circumstances.

Decline in $\mathcal{T}$ can lead to punctuated decline in the complexity of cognitive process possible within the neurodevelopmental control system, driving it into a ground state collapse in which neural systems fail to develop normally.

The essential feature is the integrated environmental insult $\hat{\rho}$. Most of the topology of the inherently unstable neurodevelopmental system has been "factored out" so that $\hat{\rho}(r_1, \ldots, r_m)$ remains the only possible index of the rate of topological information generation for the DRT. Thus, in Eqs. (4.15) and (4.16), $\mathcal{T}(\hat{\rho})$ is again the driving parameter.

Increasing $\hat{\rho}$ is then equivalent to lowering the "temperature" $\mathcal{T}$, and the system passes from high symmetry "free flow" to different forms of "crystalline" structure—broken symmetries representing the punctuated onset of significant neurodevelopmental failure.

Again, if $\kappa_2/\kappa_4 \gg \kappa_1/\kappa_3$ in Eq. (4.8), accumulated environmental insult will quickly bring the effective "temperature" below some critical value, raising the probability for, or triggering the collapse into, a dysfunctional ground state of low symmetry in which essential network connections are not made or else become locally overconnected and globally disjoint.

Sufficient conditions for the intractability—stability—of the pathological ground state can be explored using the methods of Wallace (2016). Given a vector of parameters characteristic of and driving that phase, say **J**, that measures deviations from a nonequilibrium steady state, the "free energy" analog $\mathcal{F}$ in Eq. (4.16) can be used to define a new "entropy" scalar as the Legendre transform

$$\mathcal{S} \equiv \mathcal{F}(\mathbf{J}) - \mathbf{J} \cdot \nabla_{\mathbf{J}} \mathcal{F} \tag{4.17}$$

Again, a first order dynamic equation follows using a stochastic version of the Onsager formalism from nonequilibrium thermodynamics (de Groot and Mazur 1984)

$$dJ_t^i \approx \left(\sum_k \mu_{i,k} \partial \mathcal{S} / \partial J_t^k \right) dt + \sigma_i J_t^i dB_t \tag{4.18}$$

where $\mu_{i,k}$ defines a diffusion matrix, the σ_i are parameters, and dB_t represents a noise that may be colored, i.e., not the usual Brownian motion under undifferentiated white noise.

If it is possible to factor out J^i, then Eq. (4.18) can be represented in the form

$$dJ_t^i = J_t^i dY_t^i \tag{4.19}$$

where Y_t^i is a stochastic process.

The expectation of J can then be found in terms of the Doleans-Dade exponential (Protter 1990) as

$$E(J_t^i) \propto \exp(Y_t^i - 1/2[Y_t^i, Y_t^i]) \tag{4.20}$$

where $[Y_t^i, Y_t^i]$ is the quadratic variation of the stochastic process Y_t^i (Protter 1990). Heuristically, invoking the Mean Value Theorem, if

$$1/2d[Y_t^i, Y_t^i]/dt > dY_t^i/dt \tag{4.21}$$

then the pathological ground state is stable: deviations from nonequilibrium steady state measured by J_t^i then converge in expectation to 0. That is, sufficient ongoing "noise"—determining the quadratic variation terms—can lock-in the failure of neurodevelopment with high probability, in this model.

Parallel stability arguments arise in ecosystem resilience theory (Holling 1973) which characterizes multiple quasi-stable nonequilibrium steady states among interacting populations. Pristine alpine lake ecosystems, having limited nutrient inflows, can be permanently shifted into a toxic eutrophic state by excess nutrient influx, e.g., a sewage leak, fertilizer runoff, and so on. Once shifted, the lake ecology will remain trapped in a mode of recurrent "red tide"-like plankton blooms even after sewage or fertilizer inflow is stemmed.

The quadratic variation of a stochastic process X_t, which we write as $[X_t, X_t]$, is important to understanding the pathological stability of "eutrophic" neurodevelopmental trajectories, in this model. It can be estimated from appropriate time series data using a Fourier expansion methodology adapted from financial engineering, as described in the Mathematical Appendix.

4.4 The Mitochondrial Connection

Development is not simply a matter of response to external signals, powerful as such effects may be. Metabolic free energy—the high-energy conversion of ATP to ADP—powers the many cognitive process of gene expression that must control developmental trajectories. Most directly, we can posit a Rate Distortion argument in which a developmental message is sent along biochemical channels, and the success of failure measured by complicated control and feedback mechanisms as indicated by the schematic of Fig. 4.1. In essence, there must be a parallel argument to that leading to Eq. (4.10), where $\mathcal{T}$ is replaced by the rate of metabolic free energy M.

Assuming a Gaussian channel, having the Rate Distortion Function $R(D) = (1/2)\log[\sigma^2/D]$, where D is the average distortion by the square measure (Cover and Thomas 2006) and σ^2 the inherent channel noise, we can write, for the mean of $D = \sigma^2 \exp[-2R]$,

$$< D >= \frac{\int_0^\infty \sigma^2 \exp[-2R] \exp[-R/\omega M] dR}{\int_0^\infty \exp[-R/\omega M] dR} = \frac{\sigma^2}{2\omega M + 1} \quad (4.22)$$

ω represents the efficiency with which the system converts mitochondrial free energy into control information. Small ω implies greatly increased levels of mitochondrial free energy are necessary for successful development, i.e., small $< D >$. The obvious inference is that ω will be affected by the degree of integrated environmental insult indexed by $\mathcal{T}$, so that we can, to first order at least assume $\omega = \omega_0 \mathcal{T}$ and write the synergistic relation

$$< D >= \frac{\sigma^2}{2\omega_0 \mathcal{T} M + 1} \quad (4.23)$$

inversely characterizing the success of the developmental control systems: large $< D >$ indicates failure. Other channels, as a consequence of the convexity of the Rate Distortion Function in D, will have similar expressions.

A next level of approximation takes M itself as a monotonic increasing function of $\mathcal{T}$—normalized by α_0 of Eq. (4.2)—so that, to first order, $< D > \propto 1/\mathcal{T}^2$. Under such a model, rising environmental insult, leading to the condition $\mathcal{T} < 1$, rapidly distorts developmental process by impairing mechanisms for both the generation and use of mitochondrial free energy.

4.5 Discussion and Conclusions

Chemical exposures, pre- and neonatal infections, psychosocial stress, genetic predisposition, and the cross-generational cultural and epigenetic impacts of these and other toxicants become an integrated, perhaps synergistic, signal that can overwhelm essential neurodevelopmental regulation, demanding levels of mitochondrial free energy that cannot be met. Such insult may, as well, directly interfere with the production of mitochondrial free energy. We have characterized that dynamic through statistical models based on the asymptotic limit theorems of control and information theories, models that are the functional equivalent of the usual least-squares regression based on other asymptotic limit theorems of probability theory. The greatest scientific utility of such models remains the experimental or observational comparison of similar systems under different, and different systems under similar, conditions.

The use of such tools, however, often is not easy, as the sometimes deceptive subtleties of "ordinary" regression remind us. Nonetheless, the conceptual approach taken here may still illuminate empirical studies.

A recent paper by Berman et al. (2016) describes the phenomenon of childhood-onset schizophrenia, a "pure" form of the disorder observed without the often confusing correlates of an extended disease course (Lancaster and Hall 2016). Berman et al. write:

> ...[W]e examined large-scale network interactions in childhood-onset schizophrenia...Using...resting-state functional magnetic resonance imaging...[that] identified 26 regions with decreased functional correlations in schizophrenia compared to controls...

Lancaster and Hall (2016) find that the results of Berman et al. are compatible with a pathodevelopmental model in which patients with childhood-onset schizophrenia experience excessive "over-pruning" of short-distance functional connections.

By contrast, autism spectrum disorders are marked by excessive early neural growth. Rapoport et al. (2009) assert that in autism there is an acceleration or excess of early postnatal brain development (1–3 years), whereas in childhood-onset schizophrenia (COS) there is exaggeration of the brain maturation processes of childhood and early adolescence (10–16 years):

> Both could be seen as "increased gain" of general developmental processes, albeit at different stages; both patterns could also be seen as an abnormal "shift to the left" with respect to age compared to normal brain development, with autism showing initial overgrowth and COS showing greater "pruning down" of the cortex in early and middle parts of the trajectory; both accelerations normalizing with age...

Indeed, a recent comprehensive analysis of US insurance data indicates a strong role for environmental factors in the etiology of autism spectrum disorders. Rzhetsky et al. (2014) write

> By analyzing the spatial incidence patterns of autism and intellectual disability drawn from insurance claims for nearly one third of the total US population, we found strong statistical evidence that environmental factors drive apparent spatial heterogeneity of

> both phenotypes [intellectual disability and autism] while economic incentives and population structure appear to have relatively large albeit weaker effects. The strongest predictors for autism were associated with the environment... The environmental factors implicated so far include pesticides... environmental lead..., sex hormone analogs... medications... plasticizers... and other synthetic molecules...
> It is very likely that the list of environmental factors potentially affecting development of human embryo is large and yet predominantly undocumented...
> Our results have implications for the ongoing scientific quest for the etiology of neurodevelopmental disorders. We provide evidence [for] routinely expanding the scope of inquiry to include environmental, demographic and socioeconomic factors, and governmental policies at a broad scale in a unified geospatial framework.

Environmental effects are now frequently cited as important in the etiology of autism and similar conditions (e.g., Croen et al. 2011; Landrigan 2012; DeSoto 2009). Keil and Lein (2016) in particular identify epigenetic mechanisms linking environmental chemical exposures to risk of autism spectrum disorders. Govorko et al. (2012) explore the male germline transmission of adverse effects of alcohol on fetal development.

Thus, it seems reasonable to infer that cross-generational transmission of gene methylation may also affect probabilities of ASD and other neurodevelopmental disorders. As Bohacek et al. (2013) comment,

> Psychiatric diseases are multifaceted disorders with complex etiology, recognized to have strong heritable components. Despite intense research efforts, genetic loci that substantially account for disease heritability have not yet been identified. Over the last several years, epigenetic processes have emerged as important factors for many brain diseases, and the discovery of epigenetic processes in germ cells has raised the possibility that they may contribute to disease heritability and disease risk.

They specifically note "..[E]vidence suggests that highly stressful experiences at different stages of life can markedly affect behaviors across generations and might constitute heritable risk factors for affective disorders" and go on to examine the opposite effects of chemical exposures and environmental enrichment.

One central feature of the cognitive "phase change" approach above is the possibility of a "supercooled" state during critical neurodevelopmental periods. That is, although the "temperature" defined by $\mathcal{T}$ falls below threshold for phase transition, "condensation" into nonfunctional neuronetwork configuration during a critical growth domain is made more probable rather than inevitable.

Under such a condition, however, as with supercooled liquids, some sudden perturbation can then trigger "crystallization" from high to low symmetry states, i.e., from a normal system capable of the full "global workspace" dynamics that Bernard Baars asserts are necessary and sufficient for consciousness in higher animals as described in Chap. 1, to a fractured and fragmented structure in which essential subcomponent networks are not sufficiently linked, or become, in fact, overlinked. Different condensation dynamics would broadly account for the observations of Berman et al. and Rapoport et al., the difference between autism spectrum and COS disorders being seen as different condensation phases. Typically, in such "spontaneous symmetry breaking," there will be only a small number of possible different phases. Comorbidity would be seen as the existence of both possible phase types in the same individual.

This inference may constitute the most central outcome of the modeling exercise, i.e., that "environmental" stress, in a large sense, during a critical growth regime can trigger a relatively small number of characteristic phase change analogs in neurodevelopment, although the symmetry shifts will likely involve subtle groupoid changes rather than alterations of the finite groups more familiar from network theory (Yeung 2008, Chap. 16; Golubitsky and Stewart 2006). Again, simultaneous occurrence of several such "phase condensations" would account for observed patterns of comorbidity, albeit with distinct cultural convolutions.

Indeed, as the previous chapter (following Wallace 2015c) puts it, the stabilization of human cognition via feedback from embedding social and cultural contexts is a dynamic process deeply intertwined with it, constituting the "riverbanks" directing flow of a stream of generalized consciousness at various scales and levels of organization: cultural norms and social interaction are synergistic with individual and group cognition and their disorders. That analysis finds high rates of psychopathic and antisocial personality disorder, as well as obsessive/compulsive disorder, to be culture-bound syndromes particular to Western "atomistic" societies, or to those undergoing social disintegration. Some such cultural patterning may well express itself across the forms of developmental neural malcondensation described here (e.g., Kleinman 1991; Kleinman and Cohen 1997).

While detailed application of the modeling strategies outlined here to experimental or clinical data remains to be done, the unification, after a concerted 50 year effort, of control and information theories via the Data Rate Theorem may provide opportunity for conceptual advance. Although high-end neural structures and the genetic regulators that build them are most definitely not computers in the severely limited mathematical venue of the Turing Machine, all such systems—including computers—are bounded by the asymptotic limit theorems that constrain the generation and transmission of information in the context of dynamic control.

Acknowledgements The author thanks Drs. D.N. Wallace, J.L. Rapoport, K. Keil, and P. Lein for useful discussions, and several anonymous reviewers for comments helpful in revision.

References

Atlan, H., and I. Cohen. 1998. Immune information, self-organization, and meaning. *International Immunology* 10: 711–717.

Ben-Shachar, D. 2002. Mitochondrial dysfunction in schizophrenia: A possible linkage to dopamine. *Journal of Neurochemistry* 83: 1241–1251.

Berman, R., S. Gotts, H.J. McAdams, D. Greenstein, F. Lalonde, L. Classen, R. Watsky, L. Shora, A. Ordonex, A. Raznahan, A. Martin, N. Gogtay, and J. Rapoport. 2016. Disrupted sensorimotor and social-cognitive networks underlie symptoms in childhood-onset schizophrenia. *Brain* 139: 276–291.

Black, F., and M. Scholes. 1973. The pricing of options and corporate liabilities. *Journal of Political Economy* 81: 637–654.

Bohacek, J., K. Gapp, B. Saab, and I. Mansuy. 2013. Transgenerational epigenetic effects on brain functions. *Biological Psychiatry* 73: 313–320.

Clay, H., S. Sillivan, and C. Konradi. 2011. Mitochondrial dysfunction and pathology in bipolar disorder and schizophrenia. *International Journal of Developmental Neuroscience* 29: 311–324.

Corbin, J., N. Gaiano, S. Juliano, et al. 2008. Regulation of neural progenitor cell development in the nervous system. *Journal of Neurochemistry* 106: 2272–2287.

Cover, T., and J. Thomas. 2006. *Elements of Information Theory*, 2nd ed. New York: Wiley.

Croen, L., J. Grether, C. Yoshida, R. Odouli, and V. Hendrick. 2011. Antidepressant use during pregnancy and childhood autism spectrum disorders. *Archives of General Psychiatry* 68: 1104–1112.

de Groot, S., and P. Mazur. 1984. *Nonequilibrium Thermodynamics*. New York: Dover.

DeSoto, M. 2009. Ockham's razor and autism: The case for developmental neurotoxins contributing to a disease of neurodevelopment. *Neurotoxicology* 30: 331–337.

Feynman, R. 2000. *Lectures on Computation*. Boulder CO: Westview.

Garcez, P., E.C. Loiola, R. Madeiro da Costa, et al. 2016. Zika virus impairs growth in human neurospheres and brain organoids. https://peerj.com/preprints/1817v3.

Giulivi, C., et al. 2010. Mitochondrial dysfunction in autism. *Journal of the American Medical Association* 304: 2389–2395.

Glazebrook, J.F., and R. Wallace. 2009. Rate distortion manifolds as model spaces for cognitive information. *Informatica* 33: 309–346.

Goh, S.A., Z. Dong, Y. Zhang, S. DiMauro, and B. Peterson. 2014. Mitochondrial dysfunctions as a neurobiological subtype of Autism Spectrum Disorder. *JAMA Psychiatry*. doi: 10.1001/jamapsychiatry.2014.179.

Golubitsky, M., and I. Stewart. 2006. Nonlinear dynamics of networks: The groupoid formalism. *Bulletin of the American Mathematical Society* 43: 305–364.

Govorko, D., R. Bekdash, C. Zhang, and D. Sakar. 2012. Male germline transmits fetal alcohol adverse effect on hypothalamic proopiomelanocortin gene across generations. *Biological Psychiatry* 72: 378–388.

Hayden, E. 2016. Zika highlights role of controversial fetal-tissue research: Fetal tissue may prove crucial to probing link between virus and birth defects. *Nature* 532: 16.

Holling, C. 1973. Resilience and stability of ecological systems. *Annual Reviews of Ecological Systematics* 4: 1–23.

Keil, K., and P. Lein. 2016. DNA methylation: A mechanism linking environmental chemical exposures to risk of autism spectrum disorders? *Environmental Epigenetics*. doi: 10.1093/eep/dvv012.

Kleinman, A. 1991. *Rethinking Psychiatry: From cultural category to personal experience*. New York: Free Press.

Kleinman, A., and A. Cohen. 1997. Psychiatry's global challenge. *Scientific American* 276(3): 86–89.

Lancaster, T., and J. Hall. 2016. Altered intra- inter-network dynamics reflect symptom dimensions in childhood-onset schizophrenia. *Brain* 139(10): 12.

Landrigan, P. 2010. What causes autism? Exploring the environmental contribution. *Current Opinion in Pediatrics* 22: 219–225.

Nair, G., F. Fagnani, S. Zampieri, and R. Evans. 2007. Feedback control under data rate constraints: An overview. *Proceedings of the IEEE* 95: 108–137.

Palmieri, L., and A. Persico. 2010. Mitochondrial dysfunction in autism spectrum disorders: Cause or effect? *Biochimica et Biophysica Acta* 1797: 1130–1137.

Pettini, M. 2007. *Geometry and Topology in Hamiltonian Dynamics*. New York: Springer.

Prabakaran, S., et al. 2004. Mitochondrial dysfunction in schizophrenia: Evidence for compromised brain metabolism and oxidative stress. *Molecular Psychiatry* 9: 6840697.

Protter, P. 1990. *Stochastic Integration and Differential Equations*. New York: Springer.

Rapoport, J., A. Chavez, D. Greenstein, A. Addington, and N. Gogtay. 2009. Autism-spectrum disorders and childhood onset schizophrenia: Clinical and biological contributions to a relationship revisited. *Journal of the American Academy of Childhood and Adolescent Psychiatry* 48: 10–18.

Rapoport, J., J. Giedd, and N. Gogtay. 2012. Neurodevelopmental model of schizophrenia: Update 2012. *Molecular Psychiatry* 17: 1228–1238.

Rossignol, R., and R. Frye. 2010. Mitochondrial dysfunction in autism spectrum disorders: A systematic review and meta-analysis. *Molecular Psychiatry* 17: 290–314.

Rzhetsky, A., S. Bagley, K. Wang, C. Lyttle, E. Cook, R. Altman, and R. Gibbons. 2014. Environmental and state-level regulatory factors affect the incidence of autism and intellectual disability. *PLOS Computational Biology* 10(3): e1003581.

Scaglia, F. 2010. The role of mitochondrial dysfunction in psychiatric disease. *Developmental Disabilities Research Reviews* 16: 136–143.

Shao, L., et al. 2008. Mitochondrial involvement in psychiatric disorders. *Trends in Molecular Medicine* 40: 281–295.

Tiberi, L., P. Vanderhaeghen, and J. van den Ameele. 2012. Cortical neurogenesis and morphogenesis: Diversity of cues, sources and functions. *Current Opinion in Cell Biology* 24: 269–276.

van Os, J., G. Kenis, and B. Rutten. 2010. The environment and schizophrenia. *Nature* 468: 203–212.

Wallace, R. 2012. Consciousness, crosstalk, and the mereological fallacy: An evolutionary perspective. *Physics of Life Reviews* 9: 426–453.

Wallace, R. 2015a. *An Information Approach to Mitochondrial Dysfunction: Extending Swerdlow's hypothesis*. Singapore: World Scientific.

Wallace, R. 2015b. Metabolic free energy and biological codes: A 'Data Rate Theorem' aging model. *Bulletin of Mathematical Biology* 77: 879–903.

Wallace, R. 2015c. Closed-system 'economic' models for psychiatric disorders: Western atomism and its culture-bound syndromes. *Cognitive Processing* 16: 279–290.

Wallace, R. 2016. Subtle noise structures as control signals in high-order biocognition. *Physics Letters A* 380: 726–729.

Wallace, R., and D. Wallace. 2010. *Gene Expression and Its Discontents: The Social Production of Chronic Disease*. New York: Springer.

Weinstein, A. 1996. Groupoids: Unifying internal and external symmetry. *Notices of the American Mathematical Association* 43: 744–752.

Yeung, R. 2008. *Information Theory and Network Coding*. New York: Springer.

Chapter 5
Sleep, Psychopathology, and Culture

Summary Evolutionary process has selected for inherently unstable systems in higher animals that can react swiftly to changing patterns of threat or opportunity, for example, blood pressure, the immune response, and gene expression. However, these require continual strict regulation: uncontrolled blood pressure is fatal, immune cells can attack "self" tissues, and improper gene expression triggers developmental disorders. Consciousness in particular demands high rates of metabolic free energy to both operate and regulate the fundamental biological machinery: both the "stream of consciousness" and the "riverbanks" that confine it to socially defined useful realms are constructed and reconstructed moment-by-moment in response to highly dynamic internal and environmental circumstances. We develop powerful necessary conditions models for such phenomena based on the Data Rate Theorem linking control and information theories in the context of inherent instability. The synergism between conscious action and its regulation underlies the ten-fold higher rate of metabolic energy consumption in human neural tissues and implies a close, culturally modulated relation between sleep disorders and certain psychopathologies that may be induced by environmental stress, in a large sense.

5.1 Introduction

Why do neural tissues in humans consume metabolic free energy at ten times the rate of other tissues (Clarke and Sokoloff 1999)? A simplistic answer is that, in humans, consciousness must operate with a time constant of a hundred milliseconds, and straightforward adaptation of the Arrhenius reaction rate law suggests an exponential increase of neural reaction rate with the rate of adenosine triphosphate (ATP) consumption. While this surely accounts for substantial increase in energy demand as we will argue below, an order of magnitude difference in energy consumption seems somewhat excessive. Is more going on? We will contend that, in addition to the matter of a short time constant, metabolic free energy must be supplied to parallel, rapidly operating regulatory systems that stabilize consciousness as an inherently unstable phenomenon constrained by the Data Rate Theorem, necessitating independent metabolic free energy supplies for control purposes.

R. Wallace, *Computational Psychiatry*, DOI 10.1007/978-3-319-53910-2_5

A control theory digression. At the beginning of World War II, according to report, British fighters were close adaptations of training aircraft that were inherently stable, in the sense that the aerodynamic center of pressure (CP) was well behind the machine's center of gravity (CG). Thus, hands-off, a perturbed aircraft would, after a few oscillations, return to stable flight. Early German fighters had less separation between CG and CP and were far harder to fly, but, in consequence, could turn on a dime, and were significantly better in high speed combat maneuvers than early British fighters, with serious consequences for the Allies. Current fighter jets are, by contrast, designed to be inherently unstable, thus even more highly maneuverable, and must be flown-by-wire using a number of independent computers operating under high speed majority-voting rules. For such machines, regulation is everything.

Indeed, for many physiological systems, regulation is likewise—almost—everything. Some examples.

Even our basic multicellularity seems inherently unstable: cancerous "cheating" is expected to be an ongoing threat to multi-celled organisms (Aktipis et al. 2015). Nunney (1999) looked at cancer occurrence as a function of animal size, arguing that, in larger animals, whose lifespan grows as about the 4/10 power of their cell count, prevention of cancer in rapidly proliferating tissues becomes more difficult in proportion to size. Cancer control requires the development of additional mechanisms and systems to address tumorigenesis as body size increases—a synergistic effect of cell number and organism longevity.

This pattern may represent a real barrier to the evolution of large, long-lived animals, and Nunney predicts that those that do evolve have recruited additional controls over those of smaller animals to prevent cancer. Different tissues may have evolved markedly different tumor control strategies, all energetically expensive, using different complex signaling strategies, and subject to a multiplicity of reactions to signals, including, in social animals like humans, those related to psychosocial stress.

In a similar way, gene expression itself seems inherently unstable, requiring massive regulatory machineries at every stage of growth to prevent developmental disorders. Many such stages involve critical periods in which "epigenetic" factors—including patterns of cultural input and psychosocial stress often embedded in generational history—have amplified effect (Wallace and Wallace 2009, 2010).

The immune system, seen as an independent subcomponent of the more general tumor control system (Atlan and Cohen 1998), is also inherently unstable: failure of differentiation between "self" and "nonself" leads to carcinogenic chronic inflammation (Rakoff-Nahoum 2006) and autoimmune disorders (Mackay and Rose 2014). The immune system must, then, both respond quickly to injury or pathogenic challenge and yet be closely regulated to avoid self-attack.

On a shorter timescale, unregulated blood pressure would be quickly fatal in any animal with a circulatory system. The associated baroreceptor control reflex is not simple (Rau and Elbert 2001), but can be inhibited through peripheral processes, for example, under conditions of high metabolic demand. Higher brain structures modulate the reflex, for instance, when threat is detected and fight or flight responses are being prepared. This suggests, then, that blood pressure control is a broad and actively regulated modular physiological system.

The stream of consciousness, or whatever metaphor one prefers, seems similarly regulated, and high speed mechanisms, in concert with high speed regulators, will require a high rate of metabolic free energy.

Indeed, something of this has long been known.

Schiff (2008):

> Many neurons within the central thalamus... share specific anatomical and physiological specializations that support their key role in the general functions of sustained attention, working memory, and motor preparation... Collectively, these anatomical specializations suggest that many neurons within the central thalamus may serve a general purpose function supporting large-scale cerebral dynamics associated with goal-directed behaviors and consciousness... Human imaging studies reveal that selective activation of central thalamus occurs both during tasks requiring short-term attention... and during tasks placing sustained demands of high vigilance over extended time periods...

Markman and Otto (2011):

> ...[C]haracterization of... optimality... should take into account the resource-limited nature of the human cognitive apparatus... the brain consumes a significant amount of energy. Thus energy minimization is likely to be an important constraint on cognitive processing...
>
> The ideal observer or actor defined purely in terms of information is... a point of comparison against human cognitive or sensory abilities rather than as a statement of what is optimal as a cognitive process... A definition of optimal behavior needs to take energy minimization into account.

Christie and Schrater (2015):

> [T]reating cognitive effort as a resource control problem... [implies] the brain has mechanisms that allow control of cognitive effort... [T]here are limitations on... the ability to control resource allocation... [and]... on the impact that effort has on performance... [I]t is reasonable to posit the existence of some signal by which neuronal gain is modulated according to both the availability of metabolic resource and reward signals.

Here we will attempt to formally address the deep synergistic intertwinings of metabolic energy rate, information transmission rate, and system control using powerful necessary conditions asymptotic limit theorems allowing construction of statistical models that can be fitted to data.

We will, in particular, examine the environmental induction of sleep pathologies in their inevitable cultural context.

5.2 Reaction Rate

Physiological processes such as wound healing, the immune response, tumor control, and animal consciousness all represent the evolutionary exaptation of inevitable information crosstalk into dynamic processes that recruit sets of simpler cognitive modules to build temporary working coalitions addressing particular patterns of threat and opportunity confronting an organism, as described in Chap. 1. Such tunable coalitions operate, however, at markedly different rates. Wound healing, depending on the extent of injury, may take 18 months to complete its work (Mindwood et al. 2004). Animal consciousness typically operates with a time constant of a few hundred milliseconds.

How can phenomena acting on such different rates be subsumed under the same underlying mechanism? A heuristic adaptation of Arrhenius' law (Laidler 1987), which predicts exponential differences in reaction rate with "temperature," in a large sense, produces a first approximation to the result, recognizing that, in contrast to simple chemical reactions, cognitive phenomena are inherently nonequilibrium. That is, a large class of cognitive processes can be associated with dual information sources for which palindromes are highly improbable. The easiest way to understand the association is to recognize that, at its most basic, cognition at any scale or level of organization requires choice of some particular action from a larger set of those possible. Such choice reduces uncertainty, and the reduction of uncertainty implies existence of an information source. The argument can be made quite formally, as in the first chapter.

The rate of biocognition, we nonetheless argue, appears driven by the rate of available metabolic free energy as something of a heuristic temperature analog. Absent other theory, we try the Arrhenius relation. A more different treatment, leading to similar results, can be found in Eqs. (11)–(15) and Fig. 4 of Kostal and Kobayashi (2015).

In any event, the energetics of biological reactions are remarkable: at 300 K, molecular energies represent approximately 2.5 kJ/mol in available free energy. By contrast, the basic biological energy reaction, the hydrolysis of adenosine triphosphate (ATP) to adenosine diphosphate—under proper conditions at 300 K, produces some 50 kJ/mol in reaction energy. This is equivalent to a "reaction temperature" of 6000 K. Increasing the rate of ATP delivery to one kind of tissue an order of magnitude over any others provides sufficient energy for very rapid biocognition.

The question is how such rapid biocognition is parceled out between consciousness itself and the mechanisms that must regulate and stabilize it.

In more detail, given a chemical reaction of the form $aA + bB \rightarrow pP + qQ$, the rate of change in (for example) the concentration of chemical species P (written $[P]$) is often determined by an equation like

$$d[P]/dt = \gamma(T)[A]^n[B]^m \tag{5.1}$$

where n and m depend on the reaction details. The rate constant γ is expressed by the Arrhenius relation as

$$\gamma = \alpha \exp[-E_a/CT] \tag{5.2}$$

where α is another characteristic constant, E_a is the reaction activation energy, T is the Kelvin temperature, and C a universal constant. $\exp[-E_a/CT]$ is, using the Boltzmann distribution, the fraction of molecular interactions having energy greater than E_a.

The inherently nonequilibrium nature of cognition, however, requires a slightly more sophisticated treatment.

Consciousness appears to be largely an all-or-nothing phenomenon (Sergeant and Dehaene 2004), so that conscious signal perception must exceed a threshold before becoming entrained into the characteristic general broadcast.

A direct information theory argument focuses on the Rate Distortion Function (RDF) $R(D)$ associated with the channel connecting the cognitive individual with an embedding and embodying environment. $R(D) \geq 0$, a convex function (Cover and Thomas 2006), defines the minimum rate of information transmission needed to ensure that the average distortion between what is sent and what is received is less than or equal to $D \geq 0$, according to an appropriate distortion measure. See Cover and Thomas (2006) for details. Assuming a threshold R_0 for conscious perception of an incoming signal, we can, following Feynman's (2000) identification of information as a form of free energy, write a Boltzmann-like probability for the rate of cognition as

$$P[R \geq R_0] = \frac{\int_{R_0}^{\infty} \exp[-R/\omega M]dR}{\int_0^{\infty} \exp[-R/\omega M]dR} = \exp[-R_0/\omega M] \tag{5.3}$$

where M is the supplied rate of metabolic free energy, ω a constant (representing entropic loss-in-translation), and Fig. 5.1 follows, writing $R_0/\omega = k$.

If we define an efficiency measure as (cognition rate)/M, we see that energy efficiency peaks at $M = k$, with the width of the curve depending on k; larger values generate very broad efficiency curves. The form of $\exp[-k/M]/M$ is distinct, and closely similar to what has been found in other recent work. Using numerical models of optimal coding and information transmission in Hodgkin–Huxley neurons under metabolic constraints, Kostal and Kobayashi (2015) find an almost exactly similar efficiency curve. Their treatment, however, involves regimes determined by the critical value of the effective reversal potential of an explicit neural model.

We will later interpret the efficiency peaks in Fig. 5.1 as characterizing low blood flow NREM sleep states in some higher animals.

For mammals, since body temperature remains constant, the rate of available metabolic free energy—dependent on mitochondrial function—serves as a temperature index for rates of biocognition. This determines the characteristic rate of chemically generated consciousness, or of the individual lower level cognitive modules that come together in a temporary assemblage to form such an analog. Neural tissues, in humans consuming metabolic energy at an order of magnitude greater rate than other tissues, thus can provide cognitive function many orders of magnitude faster than similar physiological phenomena.

But is this the whole story? Kostal and Kobayashi (2015) argue that efficiency matters in neural process, so that regimes of lower energy consumption may be favored over the highest cognitive rates. But, as Ristroph et al. (2013) argue, inherent instability, in itself, allows extremely rapid responses that have been strongly selected for. Here, we will argue that regulation of such phenomena must consume significant metabolic free energy, in addition to that needed for (relatively) rapid cognition.

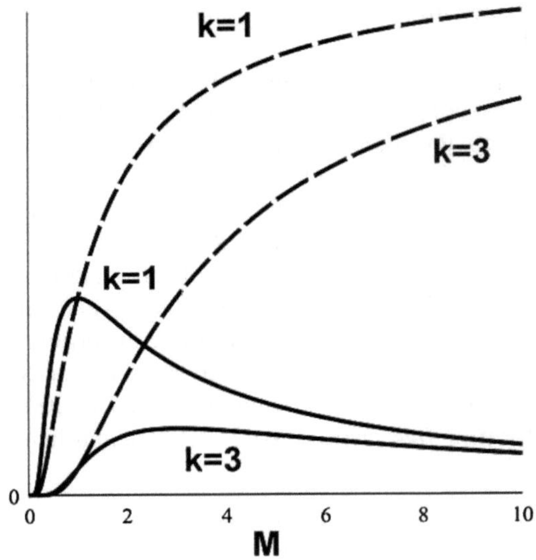

Fig. 5.1 The *dashed lines* are the "Arrhenius" relation $\exp[-k/M]$ for rate of cognition as a function of the rate of available metabolic free energy M. The *solid lines* show the efficiency measure of rate per unit metabolic energy, $\exp[-k/M]/M$. An order of magnitude increase in such free energy can enable several orders of magnitude increase in the rate of cognition. The point of greatest efficiency is at $M = k$, which we shall later interpret as an NREM sleep state in higher animals. The width of the efficiency measure, however, rapidly increases with k. Decline in M much below the shoulder of the curve triggers catastrophic collapse of cognition. The efficiency curve has the same overall form as the analogous Fig. 4 of Kostal and Kobayashi (2015), who use numerical methods

How do we understand the regulation of inherently unstable control systems? Two fundamental relations, the Data Rate and Rate Distortion Theorems, are a necessary foundation. Their convolution, we shall show, provides further insight on metabolic energy demands of high speed cognition.

5.3 The Data Rate Theorem

The Data Rate Theorem (DRT), based on an extension of the Bode Integral Theorem for linear control systems, describes the stability of feedback control under data rate constraints (Nair et al. 2007). Given a noise-free data link between a discrete linear plant and its controller, unstable modes can be stabilized only if the feedback data rate $\mathcal{H}$ is greater than the rate of "topological information" generated by the unstable system.

For the simplest incarnation—a linear expansion near a nonequilibrium steady state—the matrix equation of the plant is

$$x_{t+1} = \mathbf{A}x_t + \mathbf{B}u_t + W_t \tag{5.4}$$

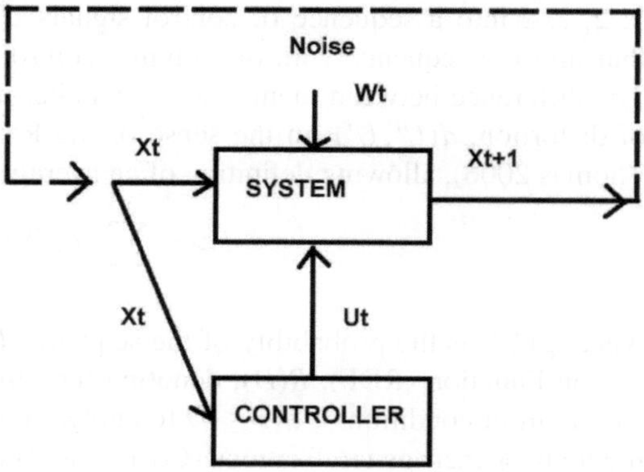

Fig. 5.2 "Regression model" for a control system near a nonequilibrium steady state. x_t is system output at time t, u_t the control signal, and W_t an added noise term

where x_t is the n-dimensional state vector at time t, u_t is the imposed n-dimensional control signal vector at time t, W_t is an added noise signal, and **A** and **B** are, in this approximation, $n \times n$ fixed matrices. See Fig. 5.2 for a schematic. Then the necessary condition for stabilizability is

$$\mathcal{H} > \log[|det(\mathbf{A}^u)|] \equiv a_0 \tag{5.5}$$

where det is the determinant and $\mathbf{A}^u$ is the decoupled unstable component of **A**, i.e., the part having eigenvalues ≥ 1. The determinant represents a generalized volume. Thus there is a critical positive data rate below which there does not exist any quantization and control scheme able to stabilize an unstable system (Nair et al. 2007).

The DRT, in its various forms, relates control theory to information theory and is as fundamental as the Shannon Coding and Source Coding Theorems, and the Rate Distortion Theorem for understanding complex biological phenomena. In the next section we derive a more biologically relevant form of the DRT than Eq. (5.5) that relates control and stability to the availability of metabolic free energy.

Accepting Feynman's (2000) insight that information is simply a form of free energy, in biological circumstances, we can write that $M = m(\mathcal{H})$, where M is the rate of metabolic free energy used to generate the control information rate $\mathcal{H}$, and m is a sharply increasing monotonic function, a consequence of massive entropic losses necessarily associated with translation of metabolic energy to information. Equation (5.5) thus implies that there is a minimum necessary rate of free energy consumption below which it is not possible to stabilize an inherently unstable biological control system. Some calculation provides details.

5.4 Rate Distortion Dynamics

We examine how the control signal u_t in Fig. 5.2 is expressed in the system response x_{t+1}. That is, we suppose it possible to deterministically retranslate—without error—a sequence of observed system outputs, $X^i = x_1^i, x_2^i \ldots$, at times

1, 2, ..., into a sequence of control signals $\hat{U}^i = \hat{u}_1^i, \hat{u}_2^i, \ldots$ and then compare that inferred sequence with the original control signal sequence $U^i = u_1^i, u_2^i, \ldots$. The difference between them is a real number measured by some chosen measure of distortion, $d(U^i, \hat{U}^i)$, in the sense of the Rate Distortion Theorem (Cover and Thomas 2006), allowing definition of an average distortion as

$$< d >= \sum_i p(U^i) d(U^i, \hat{U}^i) \tag{5.6}$$

where $p(U^i)$ is the probability of the sequence U^i. We can then define a Rate Distortion Function (RDF), $R(D)$, denoting the minimum channel capacity necessary for the limit condition $< d > \leq D$ to apply, often found using Lagrange multiplier methods or their generalizations (Cover and Thomas 2006). It is important to note that $R(D)$ is always a convex function of D, i.e., a reverse J-shaped curve, allowing deep inference (e.g., Ellis 1985).

This allows invocation of a dynamic model as follows, adopting Feynman's (2000) characterization of information as a form of free energy, measured here by the minimum channel capacity.

For a Gaussian channel, which has a Rate Distortion Function $R(D) = 1/2 \log[\sigma^2/D]$, we can now define a "Rate Distortion entropy" as the Legendre transform of the RDF as

$$S_R = R(D) - DdR(D)/dD = 1/2 \log[\sigma^2/D] + 1/2 \tag{5.7}$$

The simplest nonequilibrium Onsager equation describing the dynamics of average distortion in terms of the gradient of this "entropy" is then (de Groot and Mazur 1984)

$$dD/dt = -\mu dS_R/dD = \mu/2D \tag{5.8}$$

where t is the time and $\mu > 0$ is a diffusion coefficient. This has the solution

$$D(t) = \sqrt{\mu t} \tag{5.9}$$

which is the classic outcome of the diffusion equation. Such correspondence reduction serves as the foundation for arguing upward in both scale and complexity.

Other channels will have similar results as a consequence of the convexity of the Rate Distortion Function, i.e., monotonic growth in time. For example, a simple calculation shows that the "natural" channel, with $R(D) \propto 1/D$, gives $D(t) \propto \sqrt[3]{t}$.

Regulation, however, does not involve the diffusive drift of average distortion. Let M be the rate of metabolic free energy available for such regulation. Then a plausible model, in the presence of an internal system noise β^2 in addition to the environmental channel noise defined by σ^2, is the stochastic differential equation

$$dD_t = \left(\frac{\mu}{2D_t} - F(M)\right) dt + \frac{\beta^2}{2} D_t dW_t \tag{5.10}$$

where dW_t represents unstructured white noise and $F(M) \geq 0$ is a monotonically increasing function in the rate of metabolic free energy supply M.

This relation has the nonequilibrium steady state expectation

$$D_{nss} = \frac{\mu}{2F(M)} \tag{5.11}$$

Using the Ito chain rule on equation (5.10) (Protter 1990; Khasminskii 2012), it is possible to calculate the variance in the distortion as $E(D_t^2) - (E(D_t))^2$.

Applying the Ito relation D_t^2, however, we find that no real number solution for its expectation is possible unless the discriminant of the resulting quadratic equation is ≥ 0, giving the necessary condition

$$F(M) \geq \frac{\beta^2}{2}\sqrt{\mu} \tag{5.12}$$

Note that similar conditions will apply to other kinds of channel as a consequence of the convexity of the RDF.

From Eq. (5.11), for a Gaussian channel,

$$R_{nss} \geq \frac{1}{2}\log\left[\frac{\sigma^2\beta^2}{\sqrt{\mu}}\right] \tag{5.13}$$

Applying the Black–Scholes calculation of the Mathematical Appendix to find the "cost" of R in terms of the rate M gives

$$M = \frac{2C}{b^2}\log[R] + \kappa_1 R + \kappa_2 \tag{5.14}$$

where b is a noise term characteristic of the Black–Scholes approximation.

Substituting the result of equation (5.13) into this relation, a necessary condition for second order stability is

$$M_{nss} \geq \frac{2C}{b^2}\log\left[\frac{1}{2}\log\left[\frac{\sigma^2\beta^2}{\sqrt{\mu}}\right]\right] + \kappa_1\frac{1}{2}\log\left[\frac{\sigma^2\beta^2}{\sqrt{\mu}}\right] + \kappa_2 \equiv a_0 \tag{5.15}$$

where all constants are positive or zero.

This represents, for a Gaussian channel, another form of the Data Rate Theorem, involving an example of the iterated logarithm so characteristic of diffusion processes.

Values of M below this limit will trigger a phase transition into a disintegrated, pathological, system dynamic in a highly punctuated manner. Again, similar models can be constructed using the "natural" channel having the Rate Distortion Function $R(D) = \beta/D$.

Note, however, the two distinctly different "cost" modes implied by the conditions $\kappa_1 = 0, > 0$.

Under more complex circumstances—for example, when there are several or many possible different $F(M)$ functions representing different physiological and/or psychological states—setting the time-average expectation of dD_t in Eq. (5.10) to zero

$$< dD_t >= 0 \tag{5.16}$$

defines an *index theorem* in the sense of Hazewinkel (2002) and Atiyah and Singer (1963). An index theorem is an analytic relation whose solutions represent different topological modes of an underlying manifold, in a large sense. Such topologies are always characterized by group or groupoid structures (e.g., Lee 2000). The topologically distinct multiple solutions represent quasi-stable nonequilibrium steady state (nss) modes of the higher order maintenance system. These may range from a set of distinct nss fixed points to closed "Red Queen" cycles or pseudorandom "strange attractors" within a bounded region. Below, we will examine "directed transitions" between such modes representing large deviations in the sense of Champagnat et al. (2006), involving methods that can be applied when noise has "color."

A somewhat different attack is possible, however, having implications for the regulation and control of sleep/wake dynamics.

5.5 The Sleep Cycle: An Optimization Model

Something much like the set of quasi-stable nss states implied by the index theorem of equation (5.16)—and by the $\kappa_1 = 0, > 0$ conditions in equation (5.15)—can also be derived via an optimization argument applied to the rate calculation of equation (5.3). The essential point is that both consciousness and its necessary cognitive regulatory system(s) will follow similar metabolic scaling functions, so that we can seek to maximize a joint efficiency measure subject to constraint, applying the usual Lagrange multiplier argument. That is, letting the subscript C represent consciousness and R its regulatory machinery, we seek to maximize an efficiency functional

$$\frac{\exp[-k_C/M_C]}{M_C} + \frac{\exp[-k_R/M_R]}{M_R} \tag{5.17}$$

subject to the constraint

$$M_C + M_R = M \tag{5.18}$$

The k_X are appropriate constants and M_X is the metabolic free energy rate for process X.

Taking

$$\mathbf{\Lambda}(M_C, M_R, \lambda) \equiv \frac{\exp[-k_C/M_C]}{M_C} + \frac{\exp[-k_R/M_R]}{M_R} + \lambda(M_C + M_R - M) \tag{5.19}$$

gives the Lagrange optimization conditions as

$$\nabla_{M_C, M_R, \lambda} \mathbf{\Lambda} = 0 \tag{5.20}$$

The resulting complicated third order equation for solution pairs M_C, M_R implies the existence of several different possible optimization points for the system, strongly parameterized by the overall energy rate M. This represents another index theorem relating solutions of an analytic equation to underlying topological modes: increasing the number of systems to be optimized—adding terms of the form $\exp[-k_J/M_J]/M_J$—increases the number of possible nonequilibrium quasi-stable states.

In a first approximation to a sleep cycle model, however, there are only a small number of stages to the normal pattern: NREM (non-rapid eye movement) which involves low rates of blood flow to the brain—low M—and REM sleep which can rival or exceed conscious state blood flows. Sleep states must, like other neural processes, be highly regulated, and the argument seems to carry through, as follows.

Assuming $k_C \approx k_R$ in Eq. (5.17), direct calculation shows a symmetric efficiency curve with one or two peaks, depending on the magnitude of $M = M_C + M_R$, as in Fig. 5.3. There, taking $k_X = 1, M_C + M_R = M = 1 \rightarrow 3.6$, we obtain a maximum efficiency index, about 0.74, at $M_C = M_R \approx 1$ for $M \approx 2$, or along symmetric ridge points for larger M. The maximum value of 3.6 for M is chosen from the work of Madsen and Vorstrup (1991), who infer a 44% decline in cerebral metabolic rate during deep, slow-wave, NREM sleep.

That is, the highest efficiency peak is taken to be at the relatively quiescent NREM mode. For higher M, however, the two symmetric greatest efficiency points suggest waking/active sleep modes in which REM "sleep" may represent a parsimonious assumption of essential maintenance duties by systems otherwise dedicated to the regulation of awake consciousness. This seems analogous to the immune system which, when not extinguishing the fires of infection, wound healing, and malignancy, is deeply involved with processes of routine cellular maintenance (Cohen 2000). Long-term sleep dysfunction may be very serious indeed.

Similar ideas are in the literature. As Harris et al. (2012) put it,

> Increasingly, sleep is thought to play an energetically restorative role in the brain... [D]uring sleep there is a transient increase in ATP level in cells of awake-active regions of the brain... suggested to fuel restorative biosynthetic processes in cells that, during the day, must use all their energy on electrical and chemical signaling. This implies an energy consumption trade-off; a high use of ATP on synapses during awake periods is balanced by more ATP being allocated to other tasks during sleep.

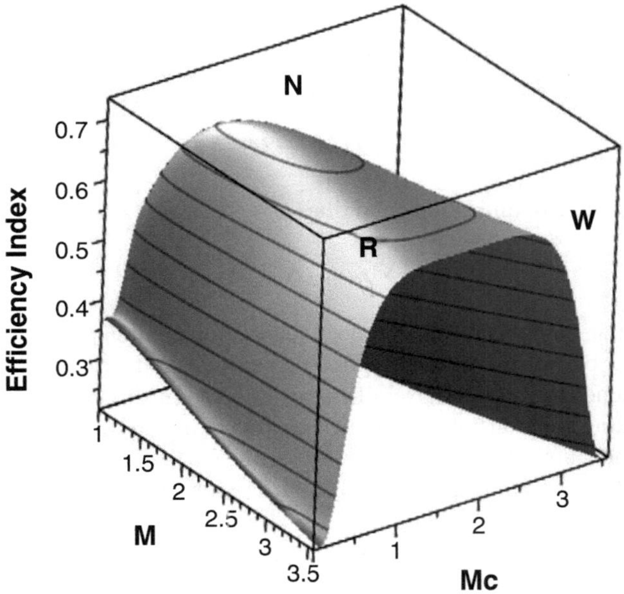

Fig. 5.3 Total consciousness-and-regulator efficiency as a function of the metabolic free energy consumed by consciousness, M_C, for $k_X = 1, M = M_C + M_R = 1 \rightarrow 3.6$. A single maximum dominates the low energy mode, suggesting NREM sleep, denoted N, representing a 44% decline from the maximum M. The maxima at higher energies are ridges, suggesting REM sleep/awake modes, denoted R, W. Normal transitions are from W to N, followed by N-R oscillation. During sleep modes, "regulator" systems for consciousness may perform routine maintenance, as does the immune system when not fighting fires of infection, wound healing, or malignancy. NREM sleep becomes a transition path between the two more active states. Changing the values of the k_X markedly shifts the relative heights and widths of the peaks, suggesting an added control mode. A little more work produces the models described in Booth and Diniz Behn (2014). There appears to be a necessary relation between sleep disturbance and the dysregulations of consciousness expressed as psychopathologies

> Energy use in the awake state also increases due to synaptic potentiation...compared to sleep...
> These changes are reversed during sleep, presumably because of homeostatic plasticity...Thus sleep is essential for adjusting synaptic energy use.

It is known that the normal route from waking, the W point in Fig. 5.3, to REM sleep, the point R, is from W to N, and then down the other ridge from N to R. Direct transition from R to W does occur, and cycling between N and R is normal.

Note that values of $k_C = k_R > 1$ generate much broader symmetric curves with far less well-defined peaks, while values less than 1 are much more sharply peaked. Unequal values raise or lower one or the other ridge. Thus "tuning" these parameters would provide significant added system control.

This model is consistent with Hobson's (2009, Fig. 4) three-dimensional Activation-Modulation-Input/Output (AIM) picture of the Western sleep cycle.

In Hobson's work, the normal progression is from waking to low-energy NREM sleep, followed by oscillation between NREM mode and the high-energy REM state.

Our model is also consistent with the observations of Hudson et al. (2014) on the punctuated recovery to consciousness from anesthesia. See their Fig. 4C which identifies three hubs between which there is directed transition during recovery.

REM sleep is now viewed as a state of consciousness which, compared to alert waking, is deficient in neuromuscular function (i.e., sleep paralysis), analytic ability, and transient episodic memory, while particularly rich in emotional cognition. In addition, both waking up and falling sleep involve complicated physiological processes to effect a transition between states. Since the late 1940s researchers have understood the importance of the reticular activating mechanism in the change from sleep to waking (Evans 2003). More recently, an arousal inhibitory mechanism, a thalamo-cortical process, has been recognized which transfers the body from waking to sleep (Evans 2003). Other work (Saper et al. 2005; Lu et al. 2006) has identified a putative neural flip-flop switch structure which controls REM sleep.

The standard phenomenological model for sleep is the "two-process" approach of Borbely (1982) and Daan et al. (1984). As Skeldon et al. (2014) put it, the homeostatic process takes the form of a relaxation oscillator that results in a monotonically increasing "sleep pressure" during wake that is dissipated during sleep. Switching from wake to sleep, and vice versa, occurs at upper and lower threshold values of the sleep pressure, respectively, where the thresholds are modulated by an approximately sinusoidal circadian oscillator. Skeldon et al. (2014) go on to show that the two-process model is essentially the same as the slow-time dynamic of the neurologically based Phillips and Robinson (2007) model.

But are there more general principles hidden here?

5.6 Transition Dynamics

How does a consciousness/regulator structure—however we choose to characterize it—make changes between the quasi-stable nss that the different modeling strategies above imply are central to the regulatory process? Indeed, recent primate experiments imply that even routine conscious decision making takes place in discrete steps (Latimer et al. 2015). Similar problems arise in evolutionary theory. Taking the approach of equation (5.10) in an evolutionary context, Champagnat et al. (2006) argue that the probability of a "large deviation" driving the system from one quasi-stable mode to another is given by a negative exponential of an entropy-like function

$$\mathcal{I} = -\sum_j P_j \log[P_j] \tag{5.21}$$

where the P_j represent a particular probability distribution. This result—the large deviations argument—is well known in numerous contexts under various names—Sanov's Theorem, the Gartner/Ellis Theorem, etc (Dembo and Zeitouni 1998).

For the composite of human consciousness-and-regulation, we argue, the transition between "states" involves the effect of impinging information sources. That is, $\mathcal{I}$ is not simply an "entropy" in this case, but represents action of an external information source (or sources) that, iteratively, regulates the internal regulators controlling individual consciousness.

A variant on this kind of approach would, for the optimization model, make different values of the essential parameters M, k_C and k_R the outputs of another, embedding, cognitive information source that drives the sleep cycle, turning off waking consciousness via the low-energy NREM mode, and then engaging high-energy REM sleep (Booth and Diniz Behn 2014).

5.7 Cultural Catalysis of the Sleep Cycle

As many have argued, humans sustain a dual heritage system of genes and culture (e.g., Richerson and Boyd 2006). Waking/REM consciousness and the associated full sleep cycle takes place in the context of a learned cultural system—an internalized information source having regularities of "grammar" and "syntax," in a large sense—that must be maintained in memory, and such maintenance requires an additional expenditure of metabolic free energy beyond what is implied by Fig. 5.3. A simple heuristic argument follows from the "chain rule" of information theory (Cover and Thomas 2006) which states that, for information sources X and Y, the joint uncertainty $H(X, Y)$ follows an inequality:

$$H(X, Y) = H(X) + H(Y|X) \leq H(X) + H(Y) \tag{5.22}$$

We can, in general, write for the probability density function of some H at a metabolic free energy rate M,

$$dP[H] = \frac{\exp[-H/kM]dH}{\int \exp[-H/kM]dH} \tag{5.23}$$

If k is very small as a consequence of an expected massive entropic loss, we can write, for the average of H,

$$< H >= \frac{\int H \exp[-H/kM]dH}{\int \exp[-H/kM]dH} \approx kM \tag{5.24}$$

so that

$$\begin{aligned} < H(X, Y) > &\approx kM_{X,Y} \leq < H(X) > + < H(Y) > \\ &\approx kM_X + kM_Y \\ M_{X,Y} &\leq M_X + M_Y \end{aligned} \tag{5.25}$$

If X represents the information source associated with the sleep/wake cycle, and Y that of internalized, learned culture, then we can expect that, through joint influence and crosstalk, culture will act as a kind of catalyst to canalize patterns of the sleep/wake cycle and their dysfunctions.

5.8 Environmental Induction of Sleep Disorders

We suppose, for the moment, that there is a scalar index of environmental stress, in a large sense, that affects patterns of sleep. Call it ρ. The stability condition of equation (5.15) then becomes something like

$$M_C(\rho) + M_R(\rho) = M(\rho) > f(\rho)\alpha_0 \tag{5.26}$$

for appropriate functional forms in ρ. We make a first order approximation, i.e., $M = \kappa_1\rho + \kappa_2, f = \kappa_3\rho + \kappa_4$, so that the stability condition becomes

$$\mathcal{T} = \frac{\kappa_1\rho + \kappa_2}{\kappa_3\rho + \kappa_4} > \alpha_0 \tag{5.27}$$

For small ρ, stability requires $\kappa_2/\kappa_4 > \alpha_0$. At high ρ the condition is $\kappa_1/\kappa_3 > \alpha_0$. If $\kappa_2/\kappa_4 \gg \kappa_1/\kappa_3 < \alpha_0$, then at some intermediate value of ρ, the basic inequality may be violated, leading to instability in the sleep/wake system. See Fig. 5.4.

Rather than a simple scalar index of stress, there are usually multiple sources that interact via different mechanisms at various scales and levels of organization. Thus we are confronted by some matrix $\hat{\rho}$ having components $\rho_{i,j} \neq \rho_{j,i}, i,j = 1,\ldots,n$ since stresses may have different effects on each other. Square matrices of order n, however, have n scalar invariants, that is, n numbers that characterize the matrix regardless of how it may be expressed in different coordinate systems. The first is the trace, the last $\pm$ the determinant. The invariants are the coefficients of the characteristic polynomial of the matrix

$$\begin{aligned}\mathcal{P}(\lambda) &= \det(\hat{\rho} - \lambda I)\\ &= \lambda^n + r_1\lambda^{n-1} + \cdots + r_{n-1}\lambda + r_n\end{aligned} \tag{5.28}$$

where λ is a parameter, det is the determinant, and I is the $n \times n$ identity matrix.

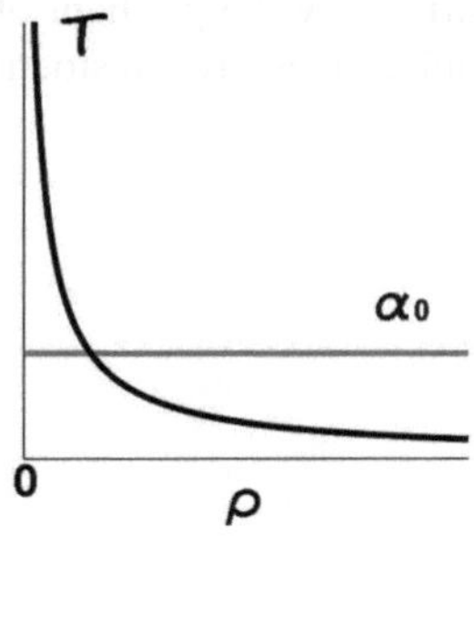

Fig. 5.4 The *horizontal line* is the limit α_0. If $\kappa_2/\kappa_4 \gg \kappa_1/\kappa_3 < \alpha_0$, at some intermediate value of integrated environmental insult ρ, $\mathcal{T} = (\kappa_1\rho + \kappa_2)/(\kappa_3\rho + \kappa_4)$ falls below criticality, and control of the sleep/wake cycle fails

We invoke the "Rate Distortion Manifold" methods of Glazebrook and Wallace (2009) or the "generalized retina" of Wallace and Wallace (2010) to argue that it is possible to accurately project high dimension information processes down onto low dimension "tangent spaces" that track an underlying complex information manifold, in a formal sense. Thus we assume there is some scalar function Γ of the matrix invariants r_i that permits definition of a $\mathcal{T}$ as in Eq. (5.27).

What are the dynamics of $\mathcal{T}(\Gamma)$ under stochastic circumstances? We again examine how a control signal u_t in Fig. 5.2 is expressed in the system response x_{t+1}. Again, we suppose it possible to deterministically retranslate a sequence of system outputs $X^i = x_1^i, x_2^i, \ldots$ into a sequence of possible control signals $\hat{U}^i = \hat{u}_0^i, \hat{u}_1^i, \ldots$ and then compare that sequence with the original control sequence $U^i = u_0^i, u_1^i, \ldots$. The difference between them is a real number measured by a chosen distortion measure d, enabling definition of an average distortion $< d >$ as in Eq. (5.6).

Again, we apply a classic Rate Distortion Theorem (RDT) argument. According to the RDT, there exists a Rate Distortion Function (RDF) that determines the minimum channel capacity, $R(D)$, necessary to keep the average distortion $< d >$ below the fixed limit D (Cover and Thomas 2006). Based on Feynman's (2000) interpretation of information as a form of (free) energy, we can now construct a Boltzmann-like probability density in the "temperature" $\mathcal{T}$ as

$$dP(R, \mathcal{T}) = \frac{\exp[-R/\mathcal{T}]dR}{\int_0^\infty \exp[-R/\mathcal{T}]dR} \tag{5.29}$$

since higher $\mathcal{T}$ necessarily implies greater channel capacity.

The integral in the denominator is essentially a statistical mechanical partition function, and we define a "free energy" Morse Function F (Pettini 2007) as

$$\exp[-F/\mathcal{T}] = \int_0^\infty \exp[-R/\mathcal{T}]dR = \mathcal{T} \tag{5.30}$$

so that $F(\mathcal{T}) = -\mathcal{T}\log[\mathcal{T}]$.

Then an entropy-analog can also be defined as the Legendre transform of F:

$$\mathcal{S} \equiv F(\mathcal{T}) - \mathcal{T}dF/d\mathcal{T} = \mathcal{T} \tag{5.31}$$

As the usual first approximation, Onsager's treatment of nonequilibrium thermodynamics (de Groot and Mazur 1984) can be applied, so that system dynamics are driven by the gradient of $\mathcal{S}$ in essential parameters—here $\mathcal{T}$—under conditions of noise. This gives a stochastic differential equation

$$\begin{aligned} d\mathcal{T}_t &\approx (\mu d\mathcal{S}/d\mathcal{T})dt + \beta\mathcal{T}_t dW_t \\ &= \mu dt + \beta\mathcal{T}_t dW_t \end{aligned} \tag{5.32}$$

where μ is a "diffusion coefficient" representing the efforts of the underlying control mechanism, and β is the magnitude of an inherent impinging white noise dW_t in the context of volatility, i.e., noise proportional to signal.

Again, via Jensen's inequality for a concave function, applying the Ito chain rule to $\log(\mathcal{T})$ in Eq. (5.32), a nonequilibrium steady state (nss) expectation for $\mathcal{T}$ can be calculated as

$$E(\mathcal{T}_t) \geq \frac{\mu}{\beta^2/2} \tag{5.33}$$

Again, μ is interpreted as indexing the attempt by the embedding control apparatus to impose stability—raise $\mathcal{T}$. Thus impinging noise can significantly increase the probability that $\mathcal{T}$ falls below the critical limit of Fig. 5.4, initiating a control failure.

However, $E(\mathcal{T})$ is an expectation, so that, in this model, there is always some nonzero probability that $\mathcal{T}$ will fall below the critical value α_0 in the multimodal expression for $\mathcal{T}(\Gamma)$: sporadic control dysfunctions have not been eliminated. Raising μ and lowering β decreases their probability, but will not drive it to zero in this model, a matter of some importance in determining population-rates of sleep disorders.

5.9 Chronic Dysfunctions of the Sleep Cycle

The normal sleep cycle involves regular transitions between the quasi-stable nonequilibrium steady states of Fig. 5.3, designated as the peak efficiency modes N, R, W. While we have described failure of the control system regulating that cycle in terms of the expectation relation of equation (5.33), the mechanism involved episodic disturbances, sporadic control dysfunctions. Here, we attempt to model chronic failure of regulation, which appears to have a somewhat different aspect.

One possible approach applies the "cognitive paradigm" of Atlan and Cohen (1998), who recognized that the immune response is not merely an automatic reflex, but involves active choice of a particular response to insult from a larger repertoire of possible responses. Choice reduces uncertainty and implies the existence of an underlying information source (Wallace 2012).

Given an information source associated with an inherently unstable, rapidly acting cognitive sleep regulation system—represented here as "dual" to it—a "natural" equivalence class algebra emerges by choosing different system origin states b_0 and defining an equivalence relation for two subsequent states b_m, b_n at times $m, n > 0$ by the existence of high probability—and hence "meaningful"—paths connecting them to the same origin point. Disjoint partition by equivalence class, similar to orbit equivalence classes in dynamical systems, defines a groupoid that is to be associated with the cognitive process. Groupoids generalize of the group concept in that there is not necessarily a product defined for each possible element pair (Weinstein 1996).

The equivalence classes define a set of cognitive dual information sources available to the inherently unstable sleep regulation system, generating a groupoid. Each orbit corresponds to a transitive groupoid whose disjoint union is the full groupoid. Each subgroupoid is associated with a distinct dual information source, and larger groupoids must have richer dual information sources than smaller.

Take X_{G_i} as the sleep cycle control system's dual information source associated with the groupoid element G_i, and Y as the information source associated with an imposed environmental stress, again in a large sense. Earlier chapters detail how the regularities of environmental exposures imply the existence of an environmental information source.

It is again possible to construct a "free energy" Morse Function (Pettini 2007). Take $H(X_{G_i}, Y) \equiv H_{G_i}$ as the joint uncertainty of the two information sources. A pseudoprobability can be written:

$$P[H_{G_i}] = \frac{\exp[-H_{G_i}/\mathcal{T}]}{\sum_j \exp[-H_{G_j}/\mathcal{T}]} \tag{5.34}$$

Again, $\mathcal{T}$ is the "temperature" from Eq. (5.27), via $\Gamma(r_1, \ldots)$, and the sum is over all possible cognitive modes.

Another Morse Function, $\mathcal{F}$, can now be defined as

$$\exp[-\mathcal{F}/\mathcal{T}] \equiv \sum_j \exp[-H_{G_j}/\mathcal{T}] \tag{5.35}$$

Since groupoids are extensions of more familiar symmetry groups, we argue by abduction that it is possible to apply an extension of Landau's picture of phase transition (Pettini 2007). In Landau's spontaneous symmetry breaking approach, phase transitions driven by temperature changes occur as alteration of system symmetry, with higher energies at higher temperatures being more symmetric. Transition between symmetries is punctuated in the temperature index $\mathcal{T}$ under the Data Rate Theorem for unstable control systems. Usually, there are only a very limited number of possible phases.

Lowering of $\mathcal{T}$ can lead to sudden, highly punctuated, decrease in the complexity of cognitive process possible within the sleep cycle control system, driving it into a kind of "ground state collapse" in which the sleep cycle is no long "normal," within expected and accepted cultural patterns.

The driving force is the integrated environmental insult Γ. Increasing Γ is then equivalent to lowering the "temperature" $\mathcal{T}$, and the system passes from high symmetry "free flow" to different forms of "crystalline" structure—broken symmetries representing the punctuated onset of regulatory failure.

To reiterate, if $\kappa_2/\kappa_4 \gg \kappa_1/\kappa_3 < a_0$ in Eq. (5.27) and its extension by replacing ρ with the composite scalar $\Gamma(r_1, \ldots)$, accumulated environmental insult will quickly bring the effective "temperature" below some critical value, raising the probability for, or triggering the collapse into, a dysfunctional ground state.

Sufficient conditions for the intractability—stability—of the pathological state can be explored as follows.

Given a vector of parameters characteristic of and driving the pathological phase, **J**, measuring deviations from a nonequilibrium steady state, the "free energy" analog $\mathcal{F}$ in Eq. (5.35) can be used to define a new "entropy" scalar as the Legendre transform

$$\mathcal{S} \equiv \mathcal{F}(\mathbf{J}) - \mathbf{J} \cdot \nabla_{\mathbf{J}} \mathcal{F} \tag{5.36}$$

As before, a first order dynamic equation follows from a stochastic version of the Onsager formalism from nonequilibrium thermodynamics (de Groot and Mazur 1984)

$$dJ_t^i \approx \left(\sum_k \mu_{i,k} \partial \mathcal{S} / \partial J_t^k \right) dt + \sigma_i J_t^i dB_t \tag{5.37}$$

where $\mu_{i,k}$ defines a diffusion matrix, the σ_i are parameters, and dB_t represents a noise that may not the usual undifferentiated white noise.

Factoring out J^i, Eq. (5.37) takes the form

$$dJ_t^i = J_t^i dY_t^i \tag{5.38}$$

where Y_t^i is a stochastic process.

The expectation of J is found in terms of the Doleans-Dade exponential (Protter 1990):

$$E(J_t^i) \propto \exp(Y_t^i - 1/2[Y_t^i, Y_t^i]) \tag{5.39}$$

taking $[Y_t^i, Y_t^i]$ as the quadratic variation of the stochastic process Y_t^i (Protter 1990). Applying the Mean Value Theorem, if

$$1/2d[Y_t^i, Y_t^i]/dt > dY_t^i/dt \tag{5.40}$$

then the (culturally sculpted) pathological ground state is stable: deviations from nonequilibrium steady state measured by J_t^i converge in expectation to 0. Thus sufficient noise—determining the quadratic variation—can lock-in the failure of the sleep cycle.

The argument is similar to that of ecosystem resilience theory (e.g., Holling 1973) characterizing multiple quasi-stable nonequilibrium steady states among interacting populations. Pristine alpine lakes, having limited nutrient inflows, can be permanently shifted into a toxic eutrophic state by excess nutrient influx—a sewage leak, fertilizer runoff, and so on. Once shifted, the lake's ecology remains trapped in a dynamic mode of recurrent "red tide"-like plankton blooms even after sewage or fertilizer inflow ceases.

5.10 Discussion and Conclusions

Evolutionary process has selected for unstable control systems in higher animals that can react relatively swiftly to patterns of threat or affordance, but require strict ongoing regulation at different scales and levels of organization. Here, we have argued that consciousness, perhaps the most significant and sophisticated rapid large-scale neural process, must be supplied with high rates of metabolic free energy to both operate and stabilize the basic physiological dynamics. That is, both the "stream of consciousness" and the "riverbanks" that confine it to socially defined useful realms are constructed and reconstructed moment-by-moment in response to highly dynamic internal and environmental circumstances. High speed response requires considerable metabolic free energy.

Further, neural structures in higher animals are "coevolutionary" in that they respond rapidly both to incoming environmental signals, in a large sense, and to signals from other neural systems. It has long been known that stabilizing coevolutionary computing systems is as inherently difficult as programming them (Wallace 2017). Although working out the full details remains to be done, punctuated transitions seem inherent.

As has often been speculated, however, failure of regulation appears to underlie many psychiatric disorders.

Emotions, Thayer and Lane (2000) assert, are an integrative index of individual adjustment to changing environmental demands, a response to an environmental event that allows rapid mobilization of multiple subsystems. Emotions allow the efficient coordination of the organism for goal-directed behavior. When the system works properly, it allows for flexible adaptation to changing environmental demands. An emotional response must be regulated to represent a proper selection of an appropriate response and the inhibition of other less appropriate responses from a more or less broad behavioral repertoire of possible responses. From their perspective, disorders of affect represent a condition in which the individual is unable to select the appropriate response, or to inhibit the inappropriate response, so that the response selection mechanism is somehow corrupted—regulation fails.

Gilbert (2001) similarly suggests that a canonical form of such corruption is the inappropriate excitation of modes that, in other circumstances, represent normal evolutionary adaptations, again representing a fundamental failure of regulation.

The formal development thus extends the perspective of Wallace (2015) on the pathologies of mitochondrial dysfunction toward realms of psychiatric disorders.

Our argument further suggest a necessarily intimate synergism between sleep disorders and a spectrum of psychopathologies. That is, if the regulatory machinery for sleep is that of consciousness off-duty, as it were, sleep disorders imply some form of consciousness failure—psychopathology. Indeed, this has long been known. As Morin and Ware (1996) put it,

> Epidemiological, cross sectional, and longitudinal data suggest a high rate of comorbidity between sleep disturbance and psychopathology, particularly between insomnia, anxiety, and depression. Between 50% and 80% of psychiatric patients complain of sleep

> disturbances during the acute phase of their illness. Conversely, among treatment-seeking individuals with a primary complaint of insomnia and randomly selected community samples, approximately one third display a concurrent psychopathology, one third exhibit psychological symptoms that do not necessarily exceed the threshold for a psychiatric disorder, and another third present insomnia as a functionally autonomous disorder. There is a positive relationship between severity of sleep disturbances and concurrent psychopathology...

More recent studies confirm the relation (e.g., Eidelman et al. 2012).

However, for humans, atomistic, individual-scale regulation must be iterated to include social and cultural influences. "Culture," to use the words of the evolutionary anthropologist Robert Boyd, "is as much a part of human biology as the enamel on our teeth," and this leads to extensions of the transition arguments above: the principal environment of humans is other humans, and we are the naked mole rats of primates. Thus social interaction and cultural heritage tend to confine individual consciousness to realms leading to socially acceptable phenotypes. Failure of such constraint is then socially constructed as misbehavior or pathology. Indeed, as described in Chap. 3, cultural norms and social interaction are generally synergistic with individual and group cognition and their disorders. Most particularly, a canonical failure mode in atomistic cultures is found to be a ground state collapse analogous to the psychopathic behaviors predicted by mainstream Western economic models. That is, high rates of psychopathic and antisocial personality disorders emerge as culture-bound syndromes characteristic of Westernizing societies, or of those undergoing large-scale social disintegration.

From the perspective of this chapter, in which consciousness-and-regulation are a synergistic composite, sleep must also reflect similar cultural influences, and this is indeed well known. That is, the structure, content, and meaning of sleep itself are very much cultural artifacts. Juillerat (1999), for example, describes sleep in the Yafar culture of Papua, New Guinea as follows:

> The conceptions [of] the Yafar about the constituents of personhood and the transformations under which they go at death... [involve] a double, the spiritual value of blood and bones, and an Ego (or soul) that becomes autonomous during sleep and communicates with the spirits of the deceased during dreams... Death does not destroy them; it reorganizes them differently. Whereas the living person experiences a temporal alternation between being awake and asleep (conscious/unconscious), this cleavage stabilizes after death, when constituents are redistributed among different categories of spirits. The quality of relations that people maintain with spirits, as well as cosmological divisions, suggest a native metapsychology.

Jeong (1995), a Korean researcher, writes

> Sleep... is a developmental product... subjected to the vicissitudes of human behavior and culture... sleep medicine/research [must involve]... a developmental perspective... Understanding of sleep and of sleep disorders is not complete without in-depth understanding of culture, philosophy, and tradition from developmental perspectives. Traditional ideas and wisdom from the past are the unavoidable resources for further understanding sleep and developing sleep research and medicine...

Worthman and Melby (2002), following a lengthy anthropological study of sleep ecologies across cultures, go further, concluding that

> ... [P]hysical, social, and temporal factors generating variation in human sleep ecology... may be paralleled by variation not only in sleep behavior but also in its physiology—specific cultural settings and practices may be associated with specific, distinctive risks for disorders of sleep and state regulation.
>
> ... [A]ssociations of cultural ecologies of sleep to such "basic" physiological regulatory systems as sleep biology, chronobiology, state regulation, and emotion regulation would imply that these systems are partially influenced or organized through cultural ecologies operating developmentally and across the life-span. Further, these associations may argue the need for attention to cultural ecology in the explanation, prevention, and possible treatment of disorders of these systems.

In sum, the composite of consciousness-and-regulation, its diurnal and other dynamics, and its pathologies, seem, for humans, as much cultural artifacts as they are biological realities. That is, an implication of our analysis is that the differences in cultural expression of waking consciousness that Nisbett et al. (2001) and other cultural psychologists have observed should be fully carried over into our understanding of all aspects of human psychology. Heine (2001) states the underlying case as follows:

> Cultural psychology does not view culture as a superficial wrapping of the self, or as a framework within which selves interact, but as something that is intrinsic to the self. It assumes that without culture there is no self, only a biological entity deprived of its potential... Cultural psychology maintains that the process of becoming a self is contingent on individuals interacting with and seizing meanings from the cultural environment...

Other higher animals, for whom culture is a less central experience, may still display dynamics somewhat similar to what we have explored here, although inferences from experiments on them may not translate well into human modalities that have been strongly sculpted by embedding culture. For humans, social interactions carrying cultural structures must impose themselves both on long-term developmental trajectories and on the conformation and dynamics of the short-term regulators that define the riverbanks confining the streams of waking and REM consciousness.

We would argue that epigenetic information sources—including, but not limited to embedding/internalized culture—act as analogs to a tunable catalyst, directing development into different characteristic pathways according to the structure of external signals, a perspective having significant implications for understanding how environmental stressors, in a large sense, can induce a broad spectrum of developmental disorders in humans.

References

Aktipis, C., A. Boddy, G. Jansen, U. Hibner, M. Hochberg. C. Maley, and G. Wilkinson. 2015. Cancer across the tree of life: Cooperation and cheating in multicellularity. *Philosophical Transactions B* 370: 20140219.

Atiyah, M., and I. Singer. 1963. The index of elliptical operators on compact manifolds. *Bulletin of the American Mathematical Society* 69: 322–433.

Atlan, H., and I. Cohen. 1998. Immune information, self-organization and meaning. *International Immunology* 10: 711–717.

Booth, V., and C. Diniz Behn. 2014. Physiologically-based modeling of sleep-wake regulatory networks. *Mathematical Biosciences* 250: 54–68.

Borbely, A. 1982. A two process model of sleep regulation. *Human Neurobiology* 1: 195–204.

Champagnat, N., R. Ferriere, and S. Meleard. 2006. Unifying evolutionary dynamics: From individual stochastic process to macroscopic models. *Theoretical Population Biology* 69: 297–321.

Christie, S.T., and P. Schrater. 2015. Cognitive cost as dynamic allocation of energetic resources. *Frontiers in Neuroscience* 9: 289 (online).

Clarke, D., and L. Sokoloff. 1999. Circulation and energy metabolism of the brain, Chapter 31. In *Basic Neurochemistry: Molecular, Cellular and Medical Aspects*, ed. G. Siegel et al., 6th ed. New York: Lippincott-Raven.

Cohen, I. 2000. *Tending Adam's Garden: Evolving the Cognitive Immune Self.* New York: Academic Press.

Cover, T., and J. Thomas. 2006. *Elements of Information Theory*. New York: Wiley.

Daan, S., D. Beersma, and A. Borbely. 1984. Timing of human sleep: Recovery process gated by a circadian pacemaker. *American Journal of Physics* 246: 161–178.

de Groot, S., and P. Mazur. 1984. *Non-equilibrium Thermodynamics*. New York: Dover Publications.

Dembo, A., and O. Zeitouni. 1998. *Large Deviations and Applications*. New York: Springer.

Eidelman, P., A. Gershon, E. McGlinchey, and A. Harvey. 2012. Sleep and psychopathology. In *The Oxford Handbook of Sleep and Sleep Disorders*, ed. G. Morin and C. Espie. New York: Oxford Handbooks.

Ellis, R. 1985. *Entropy, Large Deviations, and Statistical Mechanics*. New York: Springer.

Evans, B. 2003. Sleep, consciousness and the spontaneous and evoked electrical activity in the brain. Is there a Cortical Integrating Mechanism? *Neurophysiologie Clinique* 33: 1–10.

Feynman, R. 2000. *Lectures on Computation*. Boulder, CO: Westview Press.

Gilbert, P. 2001. Evolutionary approaches to psychopathology: The role of natural defenses. *Australian and New Zealand Journal of Psychiatry* 35: 17–27.

Glazebrook, J.F., and R. Wallace. 2009. Rate distortion manifolds as model spaces for cognitive information. *Informatica* 33: 309–345.

Harris, J., R. Jolivet, and D. Attwell. 2012. Synaptic energy use and supply. *Neuron* 75: 762–777.

Hazewinkel, M. 2002. *Encyclopedia of Mathematics, 'Index Formulas'*. New York: Springer.

Heine, S. 2001. Self as cultural object: An examination of East Asian and North American elves. *Journal of Personality* 69: 881–906.

Hobson, J.A. 2009. REM sleep and dreaming: Towards a theory of protoconsciousness. *Nature Reviews Neuroscience* 10: 803–813.

Holling, C. 1973. Resilience and stability of ecological systems. *Annual Reviews of Ecological Systematics* 4: 1–23.

Hudson, A., D. Calderon, D. Pfaff, and A. Proekt. 2014. Recovery of consciousness is mediated by a network of discrete metastable activity states. *PNAS* 111: 9283–9288.

Jeong, D.U. 1995. Sleep in Korea: A developmental perspective. *Psychiatry and Clinical Neurosciences* 49: 115–117.

Juillerat, B. 1999. Yafar death: Metamorphosis of the self in a Melanesian culture. *Homme* 149: 63–72 (in French).

Khasminskii, R. 2012. *Stochastic Stability of Differential Equations*, 2nd ed. New York: Springer.

Kostal, L., and R. Kobayashi. 2015. Optimal decoding and information transmission in Hodgkin-Huxley neurons under metabolic cost constraints. *BioSystems*. doi: 10.1016/j.biosystems.2015.06.008.

Laidler, K. 1987. *Chemical Kinetics*, 3rd ed. New York: Harper and Row.

Latimer, K., J. Yates, M. Meister, A. Huk, and J. Pillow. 2015. Single-trial spike trains in parietal cortex reveal discrete steps during decision-making. *Science* 349: 184–187.

Lee, J. 2000. *Introduction to Topological Manifolds*. New York: Springer.

Lu, J., D. Sherman, and C. Saper. 2006. A putative flip-flop switch for control of REM sleep. *Nature* 441: 589–594.

Mackay, I., and N. Rose, eds. 2014. *The Autoimmune Diseases*, 5th ed. New York: Academic Press.

Madsen, P., and S. Vorstrup. 1991. Cerebral blood flow and metabolism during sleep. *Cerebrovascular Brain Metabolism Review* 3: 281–296.

Markman, A., and A.R. Otto. 2011. Cognitive systems optimize energy rather than information. *Behavioral and Brain Sciences* 34: 207.

Mindwood, K., et al. 2004. Tissue repair and the dynamics of the extracellular matrix. *International Journal of Biochemistry and Cell Biology* 36: 1031–1037.

Morin, C., and J. Ware. 1996. Sleep and psychopathology. *Applied and Preventive Psychology* 5: 211–224.

Nair, G., F. Fagnani, S. Zampieri, and R. Evans. 2007. Feedback control under data rate constraints: An overview. *Proceedings of the IEEE* 95: 108–137.

Nisbett, R., K. Peng, C. Incheol, and A. Norenzayan. 2001. Culture and systems of thought: Holistic vs. analytic cognition. *Psychological Reviews* 108: 291–310.

Nunney, L. 1999. Lineage selection and the evolution of multistage carcinogenesis. *Proceedings of the Royal Society B* 266: 493–498.

Pettini, M. 2007. *Geometry and Topology in Hamiltonian Dynamics*. New York: Springer.

Phillips, A., and P. Robinson. 2007. A quantitative model of sleep-wake dynamics based on the physiology of the brainstem ascending arousal system. *Journal of Biological Rhythms* 22: 167–179.

Protter, P. 1990. *Stochastic Integration and Differential Equations*. New York: Springer.

Rakoff-Nahoum, S. 2006. Why cancer inflammation? *Yale Journal of Biology and Medicine* 76: 123–130.

Rau, H., and T. Elbert. 2001. Psychophysiology of arterial baroreceptors and the etiology of hypertension. *Biological Psychology* 57: 179–201.

Richerson, P., and R. Boyd. 2006. *Not by Genes Alone: How Culture Transformed Human Evolution*. Chicago: University of Chicago Press.

Ristroph, L., G. Ristroph, S. Morozova, A. Bergou, S. Chang, J. Guckenheimer, Z. Wang, and I. Cohen. 2013. Active and passive stabilization of body pitch in insect flight. *Journal of the Royal Society Interface* 10: 20130237.

Saper, C., G. Cano, and T. Scammell. 2005. Homeostatic, circadian, and emotional regulation of sleep. *Journal of Comparative Neurology* 493: 92–98.

Schiff, N. 2008. Central thalamic contributions to arousal regulation and neurological disorders of consciousness. *Annals of the New York Academy of Sciences* 1129: 105–118.

Sergeant C., and S. Dehaene. 2004. Is consciousness a gradual phenomenon? *Psychological Science* 15: 11–19.

Skeldon, A., D. Dijk, and G. Derks. 2014. Mathematical models for sleep-wake dynamics: Comparison of the two-process and a mutual inhibition neuronal model. *Plos One* 9/8/: e103877.

Thayer, J., and R. Lane. 2000. A model of neurovisceral integration in emotional regulation and dysregulation. *Journal of Affective Disorders* 61: 201–216.

Wallace, R. 2012. Consciousness, crosstalk and the mereological fallacy: An evolutionary perspective. *Physics of Life Reviews* 9: 426–453.

Wallace, R. 2015. *An Information Approach to Mitochondrial Dysfunction: Extending Swerdlow's Hypothesis*. World Scientific Series in Information Studies, vol. 4. Singapore: World Scientific.

Wallace, R. 2017. *An Information Treatment of Instabilities in Critical Systems*. Singapore: World Scientific.

Wallace, R., and D. Wallace. 2009. Code, context, and epigenetic catalysis in gene expression. *Transactions on Computational Systems Biology XI, LNABI*, vol. 5750, 283–334. Lecture Notes in Bioinformatics, Lecture Notes in Computer Science. New York: Springer.

Wallace, R., and D. Wallace. 2010. *Gene Expression and Its Discontents: The Social Production of Chronic Disease*. New York: Springer.

Weinstein, A. 1996. Groupoids: unifying internal and external symmetry. *Notices of the American Mathematical Association* 43: 744–752.

Worthman, C., and M. Melby. 2002. Toward a comparative developmental ecology of human sleep. In *Adolescent Sleep Patterns: Biological, Social, and Psychological Influences*, ed. M. Carskadon, 69–117. New York: Cambridge University Press.

Chapter 6
Embodied Cognition and Its Disorders

Summary The Data Rate Theorem that establishes a formal linkage between control theory and information theory carries deep implications for the study of embodied cognition and its dysfunctions across human, machine, and composite man–machine cockpit entities. The stabilization of such cognition is a dynamic process deeply intertwined with it, constituting, in a sense, the riverbanks directing the flow of a stream of generalized consciousness. The fundamental role of culture in human life has long been understood. Here we argue that, not only is culture, in the words of the evolutionary anthropologist Robert Boyd, "as much a part of human biology as the enamel on our teeth," but that it is an important part of the disorders of embodied cognition that different communities socially construct as mental illness. A shift in perspective on such illness is badly needed, from naive geneticism and a simplistic "biomarker" focus to a broad cognitive theory that recognizes the central roles of culture, socioeconomic structure, their histories, and their dynamics, in human development. Unfortunately, stabilizing analogs to cultural influence are unavailable to high-order real-time machine and cockpit systems that, increasingly, will be tasked with control of critical enterprises ranging from communication, power, and travel networks to nuclear and chemical reactors, organized deadly force, and so on (Wallace, Information Theory Models of Instabilities in Critical Systems. World Scientific Series in Information Studies. World Scientific, Singapore, 2017).

6.1 Introduction

Varela et al. (1991), in their pioneering study *The Embodied Mind: Cognitive Science and Human Experience*, asserted that the world is portrayed and determined by mutual interaction between the physiology of an organism, its sensorimotor circuitry, and the environment. The essential point, in their view, being the inherent structural coupling of brain–body-world. Lively debate has followed and continues (e.g., Clark 1998; Wilson 2002; Wilson and Golonka 2013). As Wilson and Golonka put it,

> The most exciting hypothesis in cognitive science right now is the theory that cognition is embodied... [T]he implications of embodiment are... radical... If cognition can span the brain, body, and environment, then the "states of mind" of disembodied cognitive science won't exist to be modified. Cognition will instead be an extended system assembled from a broad variety of resources... [that] can span brain, body, and environment...

R. Wallace, *Computational Psychiatry*, DOI 10.1007/978-3-319-53910-2_6

[We] treat perception-action problems and language problems as the same kind of thing... Linguistic information is a task resource in exactly the same way as perceptual information... Our behavior emerges from a pool of potential task resources that include the body, the environment and... the brain...

In the same year as the work by Varela et al. (1991), Brooks (1991) explored analogous ideas in robotics.

Barrett and Henzi (2005) summarize matters as follows:

...[C]ognition is "situated" and "distributed". Cognition is not limited by the "skin and skull" of the individual... but uses resources and materials in the environment... The dynamic social interactions of primates... can be investigated as cognitive processes in themselves... A distributed approach... considers all cognitive processes to emerge from the interactions between individuals, and between individuals and the world.

We will explore how this picture, of the recruitment of disparate resources into shifting, temporary coalitions in real time to address challenges and opportunities, represents a significant extension of the Baars (1988) model of consciousness, as described in Chap. 1.

It is possible to study cognition and embodiment through the Data Rate Theorem that formally relates control theory and information theory, and to include as well the needed regulation and stabilization mechanisms in a unitary construct that must interpenetrate in a similar manner.

Natural cognitive systems operate at all scales and levels of organization of biological process (e.g., Wallace 2012, 2014a). The failure of low level biological cognition in humans is often expressed through early onset of the intractable chronic diseases of senescence (e.g., Wallace and Wallace 2010, 2013; Wallace 2015). Failure of high-order cognition in humans has been the subject of intensive scientific study for over 200 years, with little if any consensus. As Johnson-Laird et al. (2006) put it,

Current knowledge about psychological illnesses is comparable to the medical understanding of epidemics in the early 19th Century. Physicians realized that cholera, for example, was a specific disease, which killed about a third of the people whom it infected. What they disagreed about was the cause, the pathology, and the communication of the disease. Similarly, most medical professionals these days realize that psychological illnesses occur... but they disagree about their cause and pathology.

As the media chatter surrounding the release of the latest official US nosology of mental disorders—the so-called DSM-V—indicates, this may be something an understatement. Indeed, the entire enterprise of the *Diagnostic and Statistical Manual of Mental Disorders* has been characterized as "prescientific" (e.g., Gilbert 2001). Atmanspacher (2006), for example, argues that formal theory of high-level cognition is itself at a point like that of physics 400 years ago, with the basic entities and the relations between them yet to be determined. Further complications arise via the overwhelming influence of culture on both mental process and its dysfunction (e.g., Heine 2001; Kleinman and Cohen 1997), something to which we will return. Previous chapters provide further exploration.

The stabilization and regulation of high-order cognition may be as complex as such cognition itself.

Some simplification, however, is possible. Cognition can be described in terms of a sophisticated real-time feedback between interior and exterior, necessarily constrained, as Dretske (1994) has noted, by certain asymptotic limit theorems of probability,

> Communication theory can be interpreted as telling one something important about the conditions that are needed for the transmission of information as ordinarily understood, about what it takes for the transmission of semantic information. This has tempted people... to exploit [information theory] in semantic and cognitive studies...
>
> ...Unless there is a statistically reliable channel of communication between [a source and a receiver]... no signal can carry semantic information... [thus] the channel over which the [semantic] signal arrives [must satisfy] the appropriate statistical constraints of information theory.

Recent intersection of that theory with the formalisms of real-time feedback systems—control theory—may provide insight into matters of embodied cognition and the parallel synergistic problem of embodied regulation and control. Here, we extend that work and apply the resulting conceptual model toward formally characterizing the unitary structural coupling of brain–body-world, a coupling in which, for humans, culture is a central mode. In the process, we will explore dynamic statistical models that can be fitted to data.

We will find, after some considerable development, results recognizably similar to those of Verduzco-Flores (2012). They, using a detailed neural network model, observed that

> [Certain network] changes result in a set of dynamics which may be associated with cognitive symptoms associated with different neuropathologies, particularly epilepsy, schizophrenia, and obsessive compulsive disorders... [S]ymptoms in these disorders may arise from similar or the same general mechanisms... [and] these pathological dynamics may form a set of overlapping states within the normal network function..[related] to observed associations between different pathologies.

Apparently, these are robust observations across a variety of cognitive systems.

6.2 The Data Rate Theorem

The Data Rate Theorem, a generalization of the Bode integral theorem for linear control systems (e.g., Yu and Mehta 2010; Kitano 2007; Csete and Doyle 2002), describes the stability of linear feedback control under data rate constraints (e.g., Mitter 2001; Tatikonda and Mitter 2004; Sahai 2004; Sahai and Mitter 2006; Minero et al. 2009; Nair et al. 2007; You and Xie 2013). Given a noise-free data link between a discrete linear plant and its controller, unstable modes can be stabilized only if the feedback data rate $\mathcal{H}$ is greater than the rate of "topological information" generated by the unstable system. For the simplest incarnation, if the linear matrix equation of the plant is of the form $x_{t+1} = \mathbf{A}x_t + \cdots$, where x_t is the n-dimensional state vector at time t, then the necessary condition for stabilizability is

$$\mathcal{H} > \log[|det\mathbf{A}^u|] \tag{6.1}$$

where det is the determinant and $\mathbf{A}^u$ is the decoupled unstable component of $\mathbf{A}$, i.e., the part having eigenvalues ≥ 1.

There is, then, a critical positive data rate below which there does not exist any quantization and control scheme able to stabilize an unstable (linear) feedback system.

This result and its variations are as fundamental as the Shannon Coding and Source Coding Theorems, and the Rate Distortion Theorem (Cover and Thomas 2006; Ash 1990; Khinchin 1957).

It is possible to entertain and extend these considerations, using methods from cognitive theory to explore brain–body-world dynamics that inherently take place under data rate constraints.

The essential analytic tool will be something much like Pettini's (2007) "topological hypothesis"—a version of Landau's spontaneous symmetry breaking insight for physical systems (Landau and Lifshitz 2007)—which infers that punctuated events often involve a change in the topology of an underlying configuration space, and the observed singularities in the measures of interest can be interpreted as a "shadow" of major topological change happening at a more basic level.

The tool for the study of such topological changes is Morse Theory (Pettini 2007; Matsumoto 2002), summarized in the Mathematical Appendix, and it is possible to construct a relevant Morse Function as a "representation" of the underlying theory.

The first step is a recapitulation of an approach to cognition using the asymptotic limit theorems of information theory (Wallace 2000, 2005a,b, 2007, 2012, 2014a).

6.3 Cognition as an Information Source

Atlan and Cohen (1998) argue that the essence of cognition involves comparison of a perceived signal with an internal, learned or inherited picture of the world, and then choice of one response from a much larger repertoire of possible responses. That is, cognitive pattern recognition-and-response proceeds by an algorithmic combination of an incoming external sensory signal with an internal ongoing activity—incorporating the internalized picture of the world—and triggering an appropriate action based on a decision that the pattern of sensory activity requires a response.

Incoming sensory input is thus mixed in an unspecified but systematic manner with internal ongoing activity to create a path of combined signals $x = (a_0, a_1, \ldots, a_n, \ldots)$. Each a_k thus represents some functional composition of the internal and the external. An application of this perspective to a standard neural network is given in Wallace (2005a, p. 34).

This path is fed into a highly nonlinear, but otherwise similarly unspecified, decision function, h, generating an output $h(x)$ that is an element of one of two disjoint sets B_0 and B_1 of possible system responses. Let

$$B_0 \equiv \{b_0, \ldots, b_k\},$$
$$B_1 \equiv \{b_{k+1}, \ldots, b_m\}.$$

Assume a graded response, supposing that if

$$h(x) \in B_0,$$

the pattern is not recognized, and if

$$h(x) \in B_1,$$

the pattern is recognized, and some action $b_j, k+1 \leq j \leq m$ takes place.

Interest focuses on paths x triggering pattern recognition-and-response: given a fixed initial state a_0, examine all possible subsequent paths x beginning with a_0 and leading to the event $h(x) \in B_1$. Thus $h(a_0, \ldots, a_j) \in B_0$ for all $0 \leq j < m$, but $h(a_0, \ldots, a_m) \in B_1$.

For each positive integer n, take $N(n)$ as the number of high probability paths of length n that begin with some particular a_0 and lead to the condition $h(x) \in B_1$. Call such paths "meaningful," assuming that $N(n)$ will be considerably less than the number of all possible paths of length n leading from a_0 to the condition $h(x) \in B_1$.

Identification of the "alphabet" of the states a_j, B_k may depend on the proper system coarse-graining in the sense of symbolic dynamics (e.g., Beck and Schlogl 1995).

Combining algorithm, the form of the function h, and the details of grammar and syntax are all unspecified in this model. The assumption permitting inference on necessary conditions constrained by the asymptotic limit theorems of information theory is that the finite limit

$$H \equiv \lim_{n \to \infty} \frac{\log[N(n)]}{n}$$

both exists and is independent of the path x. Again, $N(n)$ is the number of high probability paths of length n.

Call such a pattern recognition-and-response cognitive process ergodic. Not all cognitive processes are likely to be ergodic, implying that H, if it indeed exists at all, is path dependent, although extension to nearly ergodic processes, in a certain sense, seems possible (e.g., Wallace 2005a, pp. 31–32).

Invoking the Shannon–McMillan Theorem (Cover and Thomas 2006; Khinchin 1957), it becomes possible to define an adiabatically, piecewise stationary, ergodic information source $\mathbf{X}$ associated with stochastic variates X_j having joint and conditional probabilities $P(a_0, \ldots, a_n)$ and $P(a_n | a_0, \ldots, a_{n-1})$ such that appropriate joint and conditional Shannon uncertainties satisfy the classic relations

$$H[\mathbf{X}] = \lim_{n\to\infty} \frac{\log[N(n)]}{n}$$
$$= \lim_{n\to\infty} H(X_n|X_0,\ldots,X_{n-1})$$
$$= \lim_{n\to\infty} \frac{H(X_0,\ldots,X_n)}{n} \quad (6.2)$$

This information source is defined as dual to the underlying ergodic cognitive process, in the sense of Wallace (2005a, 2007).

"Adiabatic" means that, when the information source is properly parameterized, within continuous "pieces," changes in parameter values take place slowly enough so that the information source remains as close to stationary and ergodic as needed to make the fundamental limit theorems work. "Stationary" means that probabilities do not change in time, and "ergodic" that cross-sectional means converge to long-time averages. Between pieces it is necessary to invoke phase change formalism, a "biological" renormalization that generalizes Wilson's (1971) approach to physical phase transition (Wallace 2005a).

Shannon uncertainties $H(\ldots)$ are cross-sectional law-of-large-numbers sums of the form $-\sum_k P_k \log[P_k]$, where the P_k constitute a probability distribution. See Cover and Thomas (2006), Ash (1990), or Khinchin (1957) for the standard details.

For cognitive systems, an equivalence class algebra can be constructed by choosing different origin points a_0, and defining the equivalence of two states a_m, a_n by the existence of high probability meaningful paths connecting them to the same origin point. Disjoint partition by equivalence class, analogous to orbit equivalence classes for a dynamical system, defines the vertices of a network of cognitive dual languages that interact to actually constitute the system of interest. Each vertex then represents a different information source dual to a cognitive process. This is not a representation of a network of interacting physical systems as such, in the sense of network systems biology (e.g., Arrell and Terzic 2010). It is an abstract set of languages dual to the set of cognitive processes of interest, that may become linked into higher order structures.

Topology has become an object of algebraic study, the so-called algebraic topology, via the fundamental underlying symmetries of geometric spaces. Rotations, mirror transformations, simple ("affine") displacements, and the like uniquely characterize topological spaces, and the networks inherent to cognitive phenomena having dual information sources also have complex underlying symmetries: characterization via equivalence classes defines a groupoid, an extension of the idea of a symmetry group, as summarized by Brown (1987) and Weinstein (1996). Linkages across this set of languages occur via the groupoid generalization of Landau's spontaneous symmetry breaking arguments that will be used below (Landau and Lifshitz 2007; Pettini 2007). See the Mathematical Appendix for a brief summary of basic material on groupoids.

Recognize, however, that we are not constrained in this approach to the Atlan–Cohen model of cognition that, through the comparison with an internal picture of the world, invokes representation. The essential inference is that a broad class of cognitive phenomena—with and without representation—can be associated with a dual information source. That is, *cognition inevitably involves choice, choice reduces uncertainty, and this implies the existence of an information source.*

Again, extension to nonergodic information sources can be done using the methods of Wallace (2005a, Sect. 3.1).

6.4 Environment as an Information Source

Multifactorial cognitive and behavioral systems interact with, affect, and are affected by embedding environments that "remember" such interaction by various mechanisms. It is possible to reexpress environmental dynamics in terms of a grammar and syntax that represent the output of an information source—another generalized language.

Some examples:

1. The turn-of-the seasons in a temperate climate, for many ecosystems, looks remarkably the same year after year: the ice melts, the migrating birds return, the trees bud, the grass grows, plants and animals reproduce, high summer arrives, the foliage turns, the birds leave, frost, snow, the rivers freeze, and so on.
2. Human interactions take place within fairly well-defined social, cultural, and historical constraints, depending on context: birthday party behaviors are not the same as cocktail party behaviors in a particular social set, but both will be characteristic.
3. Gene expression during development is highly patterned by embedding environmental context via "norms of reaction," as summarized in Chap. 1.

Suppose it possible to coarse-grain the generalized "ecosystem" at time t, in the sense of symbolic dynamics (e.g., Beck and Schlogl 1995) according to some appropriate partition of the phase space in which each division A_j represent a particular range of numbers of each possible fundamental actor in the generalized ecosystem, along with associated larger system parameters. What is of particular interest is the set of longitudinal paths, system statements, in a sense, of the form $x(n) = A_0, A_1, \ldots, A_n$ defined in terms of some natural time unit of the system. Thus n corresponds to an again appropriate characteristic time unit T, so that $t = T, 2T, \ldots, nT$.

Again, the central interest is in serial correlations along paths.

Let $N(n)$ be the number of possible paths of length n that are consistent with the underlying grammar and syntax of the appropriately coarse-grained embedding ecosystem, in a large sense. As above, the fundamental assumptions are that—for this chosen coarse-graining—$N(n)$, the number of possible grammatical paths is much smaller than the total number of paths possible, and that, in the limit of (relatively) large n, $H = \lim_{n\to\infty} \log[N(n)]/n$ both exists and is independent of path.

These conditions represent a parallel with parametric statistics systems for which the assumptions are not true will require specialized approaches.

Nonetheless, not all possible ecosystem coarse-grainings are likely to work, and different such divisions, even when appropriate, might well lead to different descriptive quasi-languages for the ecosystem of interest. Thus, empirical identification of relevant coarse-grainings for which this theory will work may represent a difficult scientific problem.

Given an appropriately chosen coarse-graining, define joint and conditional probabilities for different ecosystem paths, having the form $P(A_0, A_1, \ldots, A_n)$, $P(A_n|A_0, \ldots, A_{n-1})$, such that appropriate joint and conditional Shannon uncertainties can be defined on them that satisfy Eq. (6.2).

Taking the definitions of Shannon uncertainties as above, and arguing backwards from the latter two parts of equation (6.2), it is indeed possible to recover the first, and divide the set of all possible ecosystem temporal paths into two subsets, one very small, containing the grammatically correct, and hence highly probable paths, that we will call "meaningful," and a much larger set of vanishingly low probability (Khinchin 1957).

6.5 Body Dynamics and Culture as Information Sources

Body movement is inherently constrained by evolutionary Bauplan: snakes do not brachiate, humans cannot (easily) scratch their ears with their hind legs, fish do not breathe air, nor mammals water. This is so evident that one simply does not think about it. Nonetheless, teaching a human to walk and talk, a bird to fly, or a lion to hunt, in spite of evolution, are arduous enterprises that take considerable attention from parents or even larger social groupings. Given the basic bodyplan of head and four limbs, or two feet and wings, or of a limbless spine, the essential point is that not all motions are possible. Bauplan imposes limits on dynamics. That is, if we coarse-grain motions, perhaps using some form of the standard methods for choreography transcription appropriate to the organism (or mechanism) under study, we see immediately that not all "statements" possible using the dance symbols have the same probability. That is, there will inevitably be a grammar and syntax to observed body-based behaviors imposed by evolutionary or explicit design bauplan. Sequences of symbols, say of length n, representing observed motions can be segregated into two sets, the first, and vastly larger, consisting of meaningless sequences (like humans scratching their ears with their feet) that have vanishingly small probability as $n \to \infty$. The other set, consistent with underlying bauplan grammar and syntax, can be viewed as the output of an information source, in precisely the manner of the previous two sections, in first approximation following the relations of equation (6.2).

In precisely the same manner as evolutionary Bauplan constrains possible sequences of motions into high and low probability sets, so too learned culture (and its associated patterns of social interaction) contextually constrain possible behaviors, spoken language, body postures, and many other phenotypes. That is, different cultures impose different probability structures, in a large sense, on essential matters of living and of the life course trajectory. Even sleep is widely discordant across cultural boundaries. Birth, marriage, death, social conflict, economic exchange, and so on are all strongly patterned by culture, in the context of historical trajectory and social segmentation. Some discussion of these matters in the context of mental disorder can be found in Kleinman and Good (1985), Desjarlais et al. (1995), and the references therein. Boyd and Richerson (2005) provide a more comprehensive introduction.

More generally, as Durham (1991) argues, genes and culture are two distinct but interacting systems of heritage in human populations. Information of both kinds has potential or actual influence over behaviors, creating a real and unambiguous symmetry between genes and phenotypes on the one hand, and culture and phenotypes on the other. Genes and culture are best represented as two parallel tracks of hereditary influence on phenotypes, acting, of course, on markedly different timescales. Human species' identity rests, in no small part, on its unique evolved capabilities for social mediation and cultural transmission, creating, again, high and low probability sets of real-time behavioral sequences.

6.6 Interacting Information Sources

Given a set of information sources that are linked to solve a problem, in the sense of Wilson and Golonka (2013), the "no free lunch" theorem (English 1996; Wolpert and MacReady 1995, 1997) extends a network theory-based theory (e.g., Arrell and Terzic 2010). Wolpert and Macready show there exists no generally superior computational function optimizer. That is, there is no "free lunch" in the sense that an optimizer pays for superior performance on some functions with inferior performance on others gains and losses balance precisely, and all optimizers have identical average performance. In sum, an optimizer has to pay for its superiority on one subset of functions with inferiority on the complementary subset.

This result is well known using another description. Shannon (1959) recognized a powerful duality between the properties of an information source with a distortion measure and those of a channel. This duality is enhanced if we consider channels in which there is a cost associated with the different letters. Solving this problem corresponds to finding a source that is right for the channel and the desired cost. Evaluating the Rate Distortion Function for a source corresponds to finding a channel that is just right for the source and allowed distortion level.

Yet another approach to the same result is through the "tuning theorem" of the Mathematical Appendix which inverts the Shannon Coding Theorem by noting that, formally, one can view the channel as "transmitted" by the signal. Then a dual channel capacity can be defined in terms of the channel probability distribution that maximizes information transmission assuming a fixed message probability distribution.

From the no free lunch argument, Shannon's insight, or the "Tuning Theorem", it becomes clear that different challenges facing any cognitive system, distributed collection of them, or interacting set of other information sources, that constitute an organism or automaton, must be met by different arrangements of cooperating modules represented as information sources.

It is possible to make a very abstract picture of this phenomenon based on the network of linkages between the information sources dual to the individual "unconscious" cognitive modules (UCM), and those of related information sources with which they interact. That is, a shifting, task-mapped, network of information sources is continually reexpressed: given two distinct problems classes confronting the organism or automaton, there must be two different wirings of the information sources, including those dual to the available UCM, with the network graph edges measured by the amount of information crosstalk between sets of nodes representing the different sources.

Thus fully embodied systems, in the sense of Wilson and Golonka (2013), involve interaction between very general sets of information sources assembled into a "task-specific device" in the sense of Bingham (1988) that is necessarily highly tunable. This mechanism represents a broad evolutionary generalization of the "shifting spotlight" characterizing the global neuronal workspace model of consciousness (Wallace 2005a). We will return to this point in more detail below.

The mutual information measure of crosstalk is not inherently fixed, but can continuously vary in magnitude. This suggests a parameterized renormalization: the modular network structure linked by crosstalk has a topology depending on the degree of interaction of interest.

Define an interaction parameter ω, a real positive number, and look at geometric structures defined in terms of linkages set to zero if mutual information is less than, and "renormalized" to unity if greater than, ω. Any given ω will define a regime of giant components of network elements linked by mutual information greater than or equal to it.

Now invert the argument: a given topology for the giant component will, in turn, define some critical value, ω_C, so that network elements interacting by mutual information less than that value will be unable to participate, i.e., will be locked out and not be consciously or otherwise perceived. See Chap. 1 for details. Thus ω is a tunable, syntactically dependent, detection limit that depends critically on the instantaneous topology of the giant component of linked information sources defining the analog to a global broadcast of consciousness. That topology is the basic tunable syntactic filter across the underlying modular structure, and variation in ω is only one aspect of more general topological properties that can be described in terms of index theorems, where far more general analytic constraints can become closely linked to the topological structure and dynamics of underlying networks, and, in fact, can stand in place of them (Atiyah and Singer 1963; Hazewinkel 2002).

6.7 Simple Regulation

Continuing the formal theory, information sources are often not independent, but are correlated, so that a joint information source—representing, for example, the interaction between brain, body, and the environment—can be defined having the properties

$$H(X_1, \ldots, X_n) \le \sum_{j=1}^{n} H(X_j) \tag{6.3}$$

with equality only for isolated, independent information streams.

This is the information chain rule (Cover and Thomas 2006), and has implications for free energy consumption in regulation and control of embodied cognitive processes. Feynman (2000) describes how information and free energy have an inherent duality, defining information precisely as the free energy needed to erase a message. The argument is quite direct, and it is easy to design an idealized machine that turns the information within a message directly into usable work—free energy. Information is a form of free energy and the construction and transmission of information within living things—the physical instantiation of information—consumes considerable free energy, with inevitable—and massive—losses via the second law of thermodynamics.

Suppose an intensity of available free energy is associated with each defined joint and individual information source $H(X, Y), H(X), H(Y)$, e.g., rates $M_{X,Y}$, M_X,M_Y.

Although information is a form of free energy, there is necessarily a massive entropic loss in its actual expression, so that the probability distribution of a source uncertainty H might be written in Gibbs form as

$$P[H] \approx \frac{\exp[-H/\kappa M]}{\int \exp[-H/\kappa M] dH} \tag{6.4}$$

assuming κ is very small.

To first order, then,

$$\hat{H} \equiv \int HP[H]dH \approx \kappa M \tag{6.5}$$

and, using Eq. (6.3),

$$\begin{aligned} \hat{H}(X, Y) &\le \hat{H}(X) + \hat{H}(Y) \\ M_{X,Y} &\le M_X + M_Y \end{aligned} \tag{6.6}$$

Thus, as a consequence of the information chain rule, allowing crosstalk consumes a lower rate of free energy than isolating information sources. That is, in general, it takes more free energy—higher total cost—to isolate a set of cognitive phenomena and an embedding environment than it does to allow them to engage in crosstalk.

Hence, at the free energy expense of supporting two information sources—X and Y together—it is possible to catalyze a set of joint paths defined by their joint information source. In consequence, given a cognitive module (or set of them) having an associated information source $H(\ldots)$, an external information source Y—the embedding environment—can catalyze the joint paths associated with the joint information source $H(\ldots, Y)$ so that a particular chosen developmental or behavioral pathway—in a large sense—has the lowest relative free energy.

At the expense of larger global free information expenditure—maintaining two (or more) information sources with their often considerable entropic losses instead of one—the system can feed, in a sense, the generalized physiology of a Maxwell's Demon, doing work so that environmental signals can direct system cognitive response, thus locally reducing uncertainty at the expense of larger global entropy production.

Given a cognitive biological system characterized by an information source X, in the context of—for humans—an explicitly, slowly changing, cultural "environmental" information source Y, we will be particularly interested in the joint source uncertainty defined as $H(X, Y)$, and next examine some details of how such a mutually embedded system might operate in real time, focusing on the role of rapidly changing feedback information, via the Data Rate Theorem.

6.8 Extending the Data Rate Theorem

The homology between the information source uncertainty dual to a cognitive process and the free energy density of a physical system arises, in part, from the formal similarity between their definitions in the asymptotic limit. Information source uncertainty can be defined as in the first part of equation (6.2). This is quite analogous to the free energy density of a physical system in terms of the thermodynamic limit of infinite volume (e.g., Wilson 1971; Wallace 2005a). Feynman (2000) provides a series of physical examples, based on Bennett's (1988) work, where this homology is an identity, at least for very simple systems. Bennett argues, in terms of idealized irreducibly elementary computing machines, that the information contained in a message can be viewed as the work saved by not needing to recompute what has been transmitted.

It is possible to model a cognitive system interacting with an embedding environment using an extension of the language-of-cognition approach above. Recall that cognitive processes can be formally associated with information sources, and how a formal equivalence class algebra can be constructed for a complicated cognitive system by choosing different origin points in a particular abstract "space" and defining the equivalence of two states by the existence of a high probability meaningful path connecting each of them to some defined origin point within that space.

Recall that disjoint partition by equivalence class is analogous to orbit equivalence relations for dynamical systems, and defines the vertices of a network of cognitive dual languages available to the system: each vertex represents a different information source dual to a cognitive process. The structure creates a large groupoid, with each orbit corresponding to a transitive groupoid whose disjoint union is the full groupoid, and each subgroupoid associated with its own dual information source. Larger groupoids will, in general, have "richer" dual information sources than smaller.

We can now begin to examine the relation between system cognition and the feedback of information from the rapidly changing real-time (as opposed to a slow-time cultural or other) environment, $\mathcal{H}$, in the sense of equation (6.1).

With each subgroupoid G_i of the (large) cognitive groupoid we can associate a joint information source uncertainty $H(X_{G_i}, Y) \equiv H_{G_i}$, where X is the dual information source of the cognitive phenomenon of interest, and Y that of the embedding environmental context—largely defined, for humans, in terms of culture and path-dependent historical trajectory.

Real-time dynamic responses of a cognitive system can now be represented by high probability paths connecting "initial" multivariate states to "final" configurations, across a great variety of beginning and end points. This creates a similar variety of groupoid classifications and associated dual cognitive processes in which the equivalence of two states is defined by linkages to the same beginning and end states. Thus, it becomes possible to construct a "groupoid free energy" driven by the quality of rapidly changing, real-time information coming from the embedding ecosystem, represented by the information rate $\mathcal{H}$, taken as a temperature analog.

For humans in particular, $\mathcal{H}$ is an embedding context for the underlying cognitive processes of interest, here the tunable, shifting, global broadcasts of consciousness as embedded in, and regulated by, culture. The argument-by-abduction from physical theory is, then, that $\mathcal{H}$ constitutes a kind of thermal bath for the processes of culturally channeled cognition. Thus we can, in analogy with the standard approach from physics (Pettini 2007; Landau and Lifshitz 2007), construct a Morse Function by writing a pseudoprobability for the jointly defined information sources X_{G_i}, Y having source uncertainty H_{G_i} as

$$P[H_{G_i}] = \frac{\exp[-H_{G_i}/\kappa\mathcal{H})]}{\sum_j \exp[-H_{G_j}/\kappa\mathcal{H}]} \tag{6.7}$$

where κ is an appropriate dimensionless constant characteristic of the particular system. The sum is over all possible subgroupiods of the largest available symmetry groupoid. Again, compound sources, formed by the (tunable, shifting) union of underlying transitive groupoids, being more complex, will have higher free-energy-density equivalents than those of the base transitive groupoids.

A possible Morse Function for invocation of Pettini's topological hypothesis or Landau's spontaneous symmetry breaking is then a "groupoid free energy" F defined by

$$\exp[-F/\kappa\mathcal{H}] \equiv \sum_j \exp[-H_{G_j}/\kappa\mathcal{H}] \tag{6.8}$$

It is possible, using the free energy-analog F, to apply Landau's spontaneous symmetry breaking arguments, and Pettini's topological hypothesis, to the groupoid associated with the set of dual information sources.

Many other Morse Functions might be constructed here, for example, based on representations of the cognitive groupoid(s). The resulting qualitative picture would not be significantly different. We will return to this argument below.

Again, Landau's and Pettini's insights regarding phase transitions in physical systems were that certain critical phenomena take place in the context of a significant alteration in symmetry, with one phase being far more symmetric than the other (Landau and Lifshitz 2007; Pettini 2007). A symmetry is lost in the transition—spontaneous symmetry breaking. The greatest possible set of symmetries in a physical system is that of the Hamiltonian describing its energy states. Usually states accessible at lower temperatures will lack the symmetries available at higher temperatures, so that the lower temperature phase is less symmetric: The randomization of higher temperatures ensures that higher symmetry/energy states will then be accessible to the system. The shift between symmetries is highly punctuated in the temperature index.

The essential point is that decline in the richness of the cultural control signal $\mathcal{H}$, or in the ability of that signal to influence response, as indexed by κ, can lead to punctuated decline in the complexity of cognitive process, according to this model.

This permits a Landau-analog phase transition analysis in which the quality of incoming information from the embedding regulatory system serves to raise or lower the possible richness of cognitive response to patterns of challenge. If $\kappa\mathcal{H}$ is relatively large—a rich and varied regulatory cultural environment—then there are many possible cognitive responses. If, however, noise or simple constraint limit the magnitude of $\kappa\mathcal{H}$, then behavior collapses in a highly punctuated manner to a kind of ground state in which only limited responses are possible, represented by a simplified cognitive groupoid structure.

Certain details of such information phase transitions can be calculated using "biological" renormalization methods (Wallace 2005a, Sect. 4.2) analogous to, but much different from, those used in the determination of physical phase transition universality classes (Wilson 1971).

These results represent a significant generalization of the Data Rate Theorem, as expressed in Eq. (6.1).

6.9 Another Picture

Here we use the rich vocabulary associated with the stability of stochastic differential equations to model, from another perspective, phase transitions in the composite system of "brain/body/environment" (e.g., Horsthemeke and Lefever 2006; Van den Broeck et al. 1994, 1997).

Define a "symmetry entropy" based on the Morse Function F of equation (6.8) over a set of structural parameters $\mathbf{Q} = [Q_1, \ldots, Q_n]$ (that may include $\mathcal{H}$ and other information source uncertainties) as the Legendre transform

$$\begin{aligned} S &= F(\mathbf{Q}) - \sum_i Q_i \partial F(\mathbf{Q})/\partial Q_i \\ &= F(\mathbf{Q}) - \mathbf{Q} \cdot \nabla_{\mathbf{Q}} F \end{aligned} \tag{6.9}$$

The dynamics of such a system will be driven, at least in first approximation, by Onsager-like nonequilibrium thermodynamics relations having the standard form (de Groot and Mazur 1984):

$$dQ_i/dt = \sum_j \mathcal{K}_{i,j} \partial S/\partial Q_j \tag{6.10}$$

where the $\mathcal{K}_{i,j}$ are appropriate empirical parameters and t is the time. A biological system involving the transmission of information may not have local time reversibility: in English, for example, the string "eht" has a much lower probability than "the." Without microreversibility, $\mathcal{K}_{i,j} \neq \mathcal{K}_{j,i}$.

Since, however, biological systems are quintessentially noisy, a more fitting approach is through a set of stochastic differential equations having the form

$$dQ_t^i = \mathcal{K}_i(t, \mathbf{Q})dt + \sum_j \sigma_{i,j}(t, \mathbf{Q})dB^j \tag{6.11}$$

where the $\mathcal{K}_i$ and $\sigma_{i,j}$ are appropriate functions, and different kinds of "noise" dB^j will have particular kinds of quadratic variation affecting dynamics (Protter 1990).

Several important dynamics become evident:

1. Setting the expectation of equation (6.11) equal to zero and solving for stationary points gives attractor states since the noise terms preclude unstable states. Obtaining this result, however, requires some further development.
2. This system may converge to limit cycle or pseudorandom "strange attractor" behaviors similar to thrashing in which the system seems to chase its tail endlessly within a limited venue—a kind of "Red Queen" pathology.
3. What is converged to in both cases is not a simple state or limit cycle of states. Rather it is an equivalence class, or set of them, of highly dynamic modes coupled by mutual interaction through crosstalk and other interactions. Thus "stability" in this structure represents particular patterns of ongoing dynamics rather than some identifiable static configuration or "answer." These are, then, quasi-stationary nonequilibrium states.
4. Applying Ito's chain rule for stochastic differential equations to the $(Q_t^j)^2$ and taking expectations allows calculation of variances. These may depend very powerfully on a system's defining structural constants, leading to significant instabilities depending on the magnitudes of the Q_i, as in the Data Rate Theorem (Khasminskii 2012).

5. Following the arguments of Champagnat et al. (2006), this is very much a coevolutionary structure, where fundamental dynamics are determined by the feedback between internal and external.

In particular, setting the expectation of equation (6.11) to zero generates an index theorem (Hazewinkel 2002) in the sense of Atiyah and Singer (1963), that is, an expression that relates analytic results, the solutions of the equations, to underlying topological structure, the eigenmodes of a complicated geometric operator whose groupoid spectrum represents symmetries of the possible changes that must take place for a global workspace to become activated.

6.10 Large Deviations and Epileptiform Disorders

As Champagnat et al. (2006) describe, shifts between the quasi-steady states of a coevolutionary system like that of equation (6.11) can be addressed by the large deviations formalism. The dynamics of drift away from trajectories predicted by the canonical equations can be investigated by considering the asymptotic of the probability of "rare events" for the sample paths of the diffusion.

"Rare events" are the diffusion paths drifting far away from the direct solutions of the canonical equation. The probability of such rare events is governed by a large deviation principle, driven by a "rate function" $\mathcal{I}$ that can be expressed in terms of the parameters of the diffusion.

This result can be used to study long-time behavior of the diffusion process when there are multiple attractive singularities. Under proper conditions, the most likely path followed by the diffusion when exiting a basin of attraction is the one minimizing the rate function $\mathcal{I}$ over all the appropriate trajectories.

An essential fact of large deviations theory is that the rate function $\mathcal{I}$ almost always has the canonical form

$$\mathcal{I} = -\sum_j P_j \log(P_j) \tag{6.12}$$

for some probability distribution (Dembo and Zeitouni 1998).

The argument relates to Eq. (6.11), now seen as subject to large deviations that can themselves be described as the output of an information source (or sources), say L_D, defining $\mathcal{I}$, driving Q^j-parameters that can trigger punctuated shifts between quasi-steady state topological modes of interacting cognitive submodules.

It should be clear that both internal and feedback signals, and independent, externally imposed perturbations associated with the source uncertainty $\mathcal{I}$, can cause such transitions in a highly punctuated manner. Some impacts may, in such a coevolutionary system, be highly pathological over a developmental trajectory, necessitating higher order regulatory system counterinterventions over a subsequent trajectory.

Similar ideas now pervade systems biology (Kitano 2004). See Chap. 1, Fig. 1.4.

An information source defining a large deviations rate function $\mathcal{I}$ in Eq. (6.12) can also represent input from "unexpected or unexplained internal dynamics" (UUID) unrelated to external perturbation. Such UUID will always be possible in sufficiently large cognitive systems, since crosstalk between cognitive submodules is inevitable, and any critical value can be exceeded if the structure is large enough or is driven hard enough. This suggests that, as Nunney (1999) describes for cancer, large-scale cognitive systems must be embedded in powerful regulatory structures over the life course. Wallace (2005b), in fact, examines a "cancer model" of regulatory failure for mental dysfunction.

Wallace (2000) uses a large deviations argument to examine epileptiform disorders. Martinerie et al. (1998) and Elger and Lenertz (1998) find a simplified "grammar" and "syntax" characterize brain dynamic pathways to epileptic seizure. As Martinerie et al. put it,

> The view of chronic focal epilepsy now is that abnormally discharging neurons act as pacemakers to recruit and entrain other normal neurons by loss of inhibition and synchronization into a critical mass. Thus preictal changes should be detectable during the stages of recruitment... Nonlinear indicators may undergo consistent changes around seizure onset... We demonstrated that in most cases... seizure onset could be anticipated well in advance [using nonlinear analytic methods] and that all subjects seemed to share a similar "route" towards seizure.

Wallace (2000) looks at such a phase transition in the dual source uncertainty of a spatial array of resonators itself as a large fluctuational event, having a pattern of optimal/meaningful paths defined by a (pathological) information source L_D. Failure of regulation then permits entrainment of normal neurons by abnormally discharging pacemakers, producing a seizure.

6.11 Ground State Collapse 1: Anxiety/Depression Analogs

There is, however, more to the influence of large deviations. The arguments leading to Eqs. (6.7) and (6.8) could be reexpressed using a joint information source

$$H(X_{G_i}, Y, L_D) \tag{6.13}$$

providing a more complete picture of large-scale cognitive dynamics in the presence of embedding regulatory systems, or of sporadic external "therapeutic" interventions. However, the joint information source of equation (6.13) now represents a de-facto distributed cognition involving interpenetration between both the underlying embodied cognitive process and its regulatory machinery.

That is, we can now define a composite Morse Function of embodied cognition-and-regulation, $\mathcal{F}$, as

$$\exp[-\mathcal{F}/\omega(\mathcal{H}, \mu)] \equiv \sum_i \exp[-H(X_{G_i}, Y, L_D)/\omega(\mathcal{H}, \mu)] \tag{6.14}$$

where $\omega(\mathcal{H}, \mu)$ is a monotonic increasing function of both the control data rate $\mathcal{H}$ and of the "richness" of the internal cognitive function defined by an internal—strictly cognitive—network coupling parameter μ, a more limited version of the argument in Sect. 6.6. Typical examples might include $\omega_0\sqrt{\mathcal{H}\mu}$, $\omega_0[\mathcal{H}\mu]^\gamma$, $\gamma > 0$, $\omega_1 \log[\omega_2\mathcal{H}\mu + 1]$, and so on.

More generally, $H(X_{G_i}, Y, L_D)$ in Eq. (6.14) can be replaced by the norm

$$|\Gamma_{Y,L_D}(G_i)|$$

for appropriately chosen representations Γ of the underlying cognitive-defined groupoid, in the sense of Bos (2007) and Buneci (2003). That is, many Morse Functions similarly parameterized by the monotonic functions $\omega(\mathcal{H}, \mu)$ are possible, with the underlying topology, in the sense of Pettini, itself subtly parameterized by the information sources Y and L_D.

Applying Pettini's topological hypothesis to the chosen Morse Function, reduction of either $\mathcal{H}$ or μ, or both, can trigger a "ground state collapse" representing a phase transition to a less (groupoid) symmetric "frozen" state. In higher organisms, which must generally function under real-time constraints, elaborate secondary back-up systems have evolved to take over behavioral control under such conditions. These typically range across basic emotional, as well as hypothalamic–pituitary–adrenal (HPA) and hypothalamic–pituitary–thyroid (HPT) axis, responses (e.g., Wallace 2005a, 2012, 2013; Wallace and Fullilove 2008). Failures of these systems are implicated across a vast range of common, and usually comorbid, mental and physical illnesses (e.g., Wallace 2005a,b; Wallace and Wallace 2010, 2013).

6.12 Ground State Collapse 2: Obsessive Compulsive Disorders

Following Overduin and Furnham (2012), obsessive compulsive disorder (OCD) is a widespread condition with prevalence rates from about 1% (current) and 2–2.5% (lifetime). Subclinical manifestations are frequently present in individuals without OCD, ranging perhaps as high as 25% of the general population. They state that

> Individuals with OCD, or with a high risk of developing OCD, suffer from recurrent, unwanted, and intrusive thoughts (obsessions) and engage in repetitive ritualistic behaviors (compulsions), usually aimed to prevent, reduce, or eliminate distress or feared consequences of the obsessions. Relief by rituals is generally temporary and contributes to future ritual engagement... [U]ntreated symptoms often persist or increase over time, causing significant impairment in social, professional, academic, and/or family functioning...

OCD and its subclinical manifestations thus appear widespread in studied culturally Western populations, indeed, perhaps a canonical failure mode (but see Heine 2001 for another possible interpretation).

Adapting the Onsager method of Sect. 6.9 to the Morse Function $\mathcal{F}$ of equation (6.14) leads to a generalized form of equation (6.11), now involving gradients in the extended entropy-analog

$$\mathcal{S} = \mathcal{F}(\mathbf{Q}) - \mathbf{Q} \cdot \nabla_{\mathbf{Q}} \mathcal{F} \tag{6.15}$$

Again, setting the expectation the resulting set of stochastic differential equations to zero and solving for stationary sets may provide individual nonequilibrium steady states, "Red Queen" limit cycles—where the system seems to chase its tail in repititive cycles—or even characteristic "strange attractor" sets over which the system engages in pseudorandom excursions.

Most typically, then, the Red Queen behaviors, analogous to computer thrashing, seem to provide a compelling model of OCD in sophisticated cognitive structures that are generally both embodied and distributed.

6.13 Topological Dysfunctions: Autism Spectrum and Schizophreniform Analogs

Recall that the global workspace model of consciousness (Baars 1988; Baars et al. 2013; Wallace 2005a, 2007, 2012) posits a "theater spotlight" involving the recruitment of unconscious cognitive modules of the brain into a temporary, tunable, general broadcast fueled by crosstalk that allows formation of the shifting coalitions needed to address real-time problems facing a higher organism. Similar exaptations of crosstalk between cognitive modules at smaller scales have been recognized in wound healing, the immune system, and so on (Wallace 2012, 2014a). Theories of embodied cognition envision that phenomenon as analogous, that is, as the temporary assembly of interacting modules from brain, body, and environment to address real-time problems facing an organism. This is likewise a dynamic process that sees many available information sources—not limited to those dual to cognitive brain or internal physiological modules—again linked by crosstalk into a tunable real-time phenomenon that is, in effect, a generalized consciousness.

The perspective has particular implications for disorders of brain connectivity like schizophrenia and autism. Figure 6.1 shows a schematic of a "generalized consciousness" involving dynamic patterns of crosstalk between information sources—the X_j—representing brain, body, culture, socioeconomy, and environment, in no particular order, and treated as fundamentally equivalent. The full and dotted lines represent recruitment of these dispersed resources (involving crosstalk at or above some tunable value ω) in two different topological patterns to address two different kinds of problems in real time.

"Mental disorders," in a large sense, emerge as a synergistic dysfunction of internal process and regulatory milieu, which above was simply characterized by the interaction between the driving parameters μ and $\mathcal{H}$. Other forms of

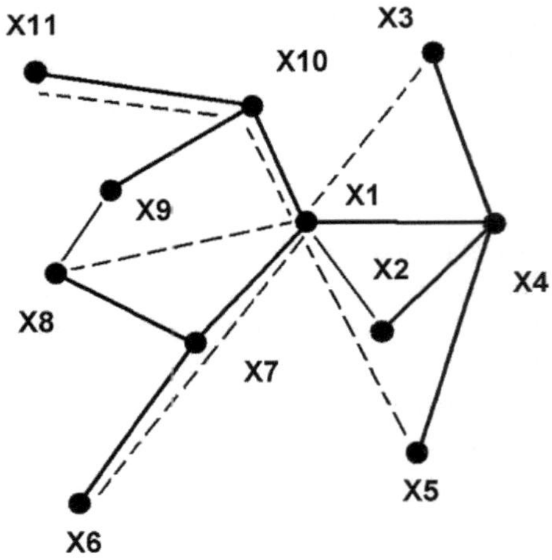

Fig. 6.1 *Full and dotted lines* represent two different recruitments of brain, body, cultural, socioeconomic, and other environmental information sources to address real-time problems facing an individual (or composite distributed cognition system). Both underlying topology and the crosstalk index $\omega(\mathcal{H}, \mu)$ are dynamically tunable, representing a generalized consciousness. Pathological restrictions on connectivity or topology would be manifest as analogs to autism or schizophrenia, in this model, in addition to the "anxiety/depression" mode of ground state collapse

dysfunction likely involve characteristic irregularities in topological connections. For example, autism spectrum and schizophreniform disorders are widely viewed as caused by failures in linkage that limit recruitment of unconscious cognitive brain modules (e.g., Wallace 2005b). Thus analogous disorders might arise from similar "topological failures" affecting the real-time recruitment of brain, body, culture, regulatory, and environmental information sources. The central role of culture in human biology means, of course, that, for humans, all such disorders are inherently "culture bound syndromes," much in the spirit of Kleinman and Cohen (1997) and Heine (2001).

6.14 Discussion and Conclusions

Here, we have made formal use of the Data Rate Theorem in exploring the dynamics of such an embodied cognition, and of a necessarily related embodied regulation. These, according to theory, inevitably involve a synergistic interpenetration among nested sets of actors, represented here as information sources. They may include dual sources to internal cognitive modules, body bauplan, environmental information, language, culture, and so on. Following the arguments of Wallace (2014b), similar considerations apply to machine and man/machine cockpit systems, a matter discussed more fully in Wallace (2017) and addressed in Chaps. 8 and 9.

Two factors determine the possible range of real-time cognitive response, in the simplest version of the model. These are the magnitude of the environmental feedback signal and the inherent structural richness of the underlying cognitive groupoid. If that richness is lacking—if the possibility of internal μ-connections is limited—then even very high levels of $\mathcal{H}$ may not be adequate to activate appropriate behavioral responses to important real-time feedback signals, following the argument of equation (6.14).

Cognition and regulation must, then, be viewed as interacting gestalt processes, involving not just an atomized individual (or, taking an even more limited "NIMH" perspective, just the brain of that individual), but the individual in a rich context that must include both the body that acts on the environment, and the environment that reacts on body and brain. Huys et al. (2016) make much the same point.

The large deviations analysis suggests that cognitive function also occurs in the context, not only of a powerful environmental embedding, but also of a specific regulatory milieu: there can be no cognition without regulation. The "stream of generalized consciousness" represented by embodied cognition must be contained within regulatory riverbanks.

For humans and some other animal species (e.g., Avital and Jablonka 2000), this view must be expanded by another layer of information sources: as the evolutionary anthropologist Robert Boyd has expressed it, "Culture is as much a part of human biology as the enamel on our teeth." Thus, for humans, the schematic hierarchy of interacting information sources becomes

$$\textbf{Brain} \rightarrow \textbf{Body} \rightarrow \textbf{Culture} \rightarrow \textbf{Environment}$$

Current theory on embodied cognition omits the critical level of cultural modulation, which is, by structure, unavailable to machine and man/machine systems.

We thus significantly extend the criticisms of Bennett and Hacker (2003) who examined the mereological fallacy of a decontextualization that attributes to "the brain" what is the province of the whole individual. Here, the "whole individual" includes essential interactions with embedding environmental and regulatory settings that, for humans, must include cultural heritage and socioeconomic dynamics.

We have explored in particular epileptiform seizures and "ground state collapses" analogous to anxiety/depression and OCD, but more diverse and subtle "topological" failures seem likely. As Johnson-Laird et al. (2006) indicate, surprisingly little is known about such dysfunction in humans. For some time, the study of mental disorders has been strongly dominated by a simplistic paradigm driven largely by the interests of the pharmaceutical industry, which has since abandoned the effort as a dry hole. The story is well known, and parallels the arguments in Chap. 1 of Wallace and Wallace (2013).

A shift in perspective is needed to a comprehensive cognitive theory of mental disorders that recognizes the central roles of culture, socioeconomic structure, and their dynamics in both healthy human development and its pathologies. At present, little such work is actively supported in the USA, for deep cultural, ideological, and political reasons.

In addition, this work can be seen as extending the considerations of Wallace (2014b, 2017) from a simple "ground state collapse" of machines and man/machine cockpits in which "all possible targets are enemies" to spectra of more subtle failures that will afflict real-time systems increasingly being given control of critical human structures and processes. Indeed, considering the collapse of "pharmaceutical industry" paradigms, it seems possible that study of machine and man/machine failures could provide the intellectual and financial horsepower necessary to lift psychiatry from its longstanding and ideologically enforced doldrums.

References

Arrell, D., and A. Terzic. 2010. Network systems biology for drug discovery. *Clinical Pharmacology and Therapeutics* 88: 120–125.
Ash, R. 1990. *Information Theory*. New York: Dover Publications.
Atiyah, M., and I. Singer. 1963. The index of elliptical operators on compact manifolds. *Bulletin of the American Mathematical Society* 69: 322–433.
Atlan, H., and I. Cohen. 1998. Immune information, self-organization, and meaning. *International Immunology* 10: 711–717.
Atmanspacher, H. 2006. Toward an information theoretical implementation of contextual conditions for consciousness. *Acta Biotheoretica* 54: 157–160.
Avital, E., and E. Jablonka. 2000. *Animal Traditions*. New York: Cambridge University Press.
Baars, B. 1988. *A Cognitive Theory of Consciousness*. New York: Cambridge University Press.
Baars, B., S. Franklin, and T. Ramsoy. 2013. Global workspace dynamics: Cortical 'binding and propagation' enables conscious contents. *Frontiers in Psychology* 4: Artile 200.
Barrett, L., and P. Henzi. 2005. The social nature of primate cognition. *Proceedings of the Royal Society B* 272: 1865–1875.
Beck, C., and F. Schlogl. 1995. *Thermodynamics of Chaotic Systems*. New York: Cambridge University Press.
Bennett, C. 1988. Logical depth and physical complexity. In *The Universal Turing Machine: A Half-Century Survey*, ed. R. Herkin, 227–257. New York: Oxford University Press.
Bennett, M., and P. Hacker. 2003. *Philosophical Foundations of Neuroscience*. London: Blackwell Publishing.
Bingham, G. 1988. Task-specific devices and the perceptual bottleneck. *Human Movement Science* 7: 225–264.
Bos, R. 2007. Continuous representations of groupoids. Arxiv:math/0612639.
Boyd, R., and P. Richerson. 2005. *The Origin and Evolution of Cultures*. New York: Oxford University Press.
Brooks, R. 1991. Intelligence without representation. *Artificial Intelligence* 47: 139–159.
Brown, R. 1987. From groups to groupoids: a brief survey. *Bulletin of the London Mathematical Society* 19: 113–134.
Buneci, M. 2003. *Representare de Groupoizi*. Timisoara: Editura Mirton.
Champagnat, N., R. Ferrier, and S. Meleard. 2006. Unifying evolutionary dynamics: From individual stochastic process to macroscopic models. *Theoretical Population Biology* 69: 297–321.
Clark, A. 1998. Embodied, situated and distributed cognition. In *A Companion to Cognitive Science*, ed. W. Bechtal and G. Graham, 506–517. London: Blackwell.
Cover, T., and J. Thomas. 2006. *Elements of Information Theory*, 2nd ed. New York: Wiley.
Csete, M., and J. Doyle. 2002. Reverse engineering of biological complexity. *Science* 295: 1664–1669.

de Groot, S., and P. Mazur, 1984. *Nonequilibrium Thermodynamics*. New York: Dover.
Dembo, A., and O. Zeitouni. 1998. *Large Deviations: Techniques and Applications*, 2nd ed. New York: Springer.
Desjarlais, R., et al. 1995. *World Mental Health*. New York: Oxford University Press.
Dretske, F. 1994. The explanatory role of information. *Philosophical Transactions of the Royal Society A* 349: 59–70.
Durham, W. 1991. *Coevolution: Genes, Culture, and Human Diversity*. Menlo Park, CA: Stanford University Press.
Elger, C., and K. Lenertz. 1998. Seizure prediction by nonlinear time series analysis of brain electrical activity. *European Journal of Neuroscience* 10: 786–789.
English, T. 1996. Evaluation of evolutionary and genetic optimizers: No free lunch. In *Evolutionary Programming V: Proceedings of the Fifth Annual Conference on Evolutionary Programming*, ed. L. Fogel, P. Angeline, and T. Back, 163–169. Cambridge, MA: MIT Press.
Feynman, R. 2000. *Lectures on Computation*. New York: Westview Press.
Gilbert, P. 2001. Evolutionary approaches to psychopathology: The role of natural defenses, Australian and New Zealand. *Journal of Psychiatry* 35: 17–27.
Hazewinkel, M. 2002. *Index Formulas, Encyclopedia of Mathematics*. New York: Springer.
Heine, S. 2001. Self as cultural product: An examination of East Asian and North American selves. *Journal of Personality* 69: 881–906.
Horsthemeke, W., and R. Lefever. 2006. Noise-induced Transitions. In *Theory and Applications in Physics, Chemistry, and Biology*, vol. 15. New York: Springer.
Huys, Q., T. Maia, and M. Frank. 2016. Computational psychiatry as a bridge from neuroscience to clinical applications. *Nature Neuroscience* 19: 404–413.
Johnson-Laird, P., F. Mancini, and A. Gangemi. 2006. A hyperemotional theory of psychological illness. *Psychological Review* 113: 822–841.
Kitano, A. 2007. Towards a theory of biological robustness. *Molecular Systems Biology*, 3: 137 (online).
Khasminskii, R. 2012. *Stochastic Stability of Differential Equations*. New York: Springer.
Khinchin, A. 1957. *The Mathematical Foundations of Information Theory*. New York: Dover Publications.
Kitano, H. 2004. Biological robustness. *Nature Genetics* 5: 826–837.
Kleinman, A., and A. Cohen. 1997. Psychiatry's global challenge. *Scientific American* 276(3): 86–89.
Kleinman, A., and B. Good. 1985. *Culture and Depression*. San Francisco: California University Press.
Landau, L., and E. Lifshitz. 2007. *Statistical Physics*, 3rd ed. Part I. New York: Elsevier.
Martinerie, J., C. Adam, and M. Le Van Quyen. 1998. *Nature Medicine* 4: 1173–1176.
Matsumoto, Y. 2002. *An Introduction to Morse Theory*. Providence, RI: American Mathematical Society.
Minero, P., M. Franceschetti, S. Dey, and G. Nair. 2009. Data Rate Theorem for stabilization over time-varying feedback channels. *IEEE Transactions on Automatic Control* 54: 243–255.
Mitter, S. 2001. Control with limited information. *European Journal of Control* 7: 122–131.
Nair, G., F. Fagnani, S. Zampieri, and R. Evans. 2007. Feedback control under data rate constraints: An overview. *Proceedings of the IEEE* 95: 108–137.
Nunney, L. 1999. Lineage selection and the evolution of multistage carcinogenesis. *Proceedings of the London Royal Society B* 266: 493–498.
Overduin, M., and A. Furnham. 2012. Assessing obsessive-compulsive disorder (OCD): A review of self-report measures. *Journal of Obsessive-Compulsive and Related Disorders* 1: 312–324.
Pettini, M. 2007. *Geometry and Topology in Hamiltonian Dynamics and Statistical Mechanics*. New York: Springer.
Protter, P. 1990. *Stochastic Integration and Differential Equations*. New York: Springer.
Sahai, A. 2004. The necessity and sufficiency of anytime capacity for control over a noisy communication link, decision and control. In *43rd IEEE Conference on CDC*, vol. 2, 1896–1901.
Sahai, A., and S. Mitter. 2006. The necessity and sufficiency of anytime capacity for control over a noisy communication link Part II: Vector systems. http://arxiv.org/abs/cs/0610146.

Shannon, C. 1959. Coding theorems for a discrete source with a fidelity criterion. *Institute of Radio Engineers International Convention Record* 7: 142–163.
Tatikonda, S., and S. Mitter. 2004. Control over noisy channels. *IEEE Transactions on Automatic Control* 49: 1196–1201.
Van den Broeck, C., J. Parrondo, and R. Toral. 1994. Noise-induced nonequilibrium phase transition. *Physical Review Letters* 73: 3395–3398.
Van den Broeck, C., J. Parrondo, R. Toral, and R. Kawai. 1997. Nonequilibrium phase transitions induced by multiplicative noise. *Physical Review E* 55: 4084–4094.
Varela, F., E. Thompson, and E. Rosch. 1991. *The Embodied Mind: Cognitive Science and Human Experience*. Cambridge, MA: MIT Press.
Verduzco-Flores, S., B. Ermentrout, and M. Bodner. 2012. Modeling neuropathologies as disruption of normal sequence generation in working memory models. *Neural Networks* 27: 21–31.
Wallace, R. 2000. Language and coherent neural amplification in hierarchical systems: Renormalization and the and the dual information source of a generalized spatiotemporal stochastic resonance. *International Journal of Bifurcation and Chaos* 10: 493–502.
Wallace, R. 2005a. *Consciousness: A Mathematical Treatment of the Global Neuronal Workspace Model*. New York: Springer.
Wallace, R. 2005b. A global workspace perspective on mental disorders. *Theoretical Biology and Medical Modelling* 2: 49.
Wallace, R. 2007. Culture and inattentional blindness: A global workspace perspective. *Journal of Theoretical Biology* 245: 378–390.
Wallace, R. 2012. Consciousness, crosstalk, and the mereological fallacy: An evolutionary perspective. *Physics of Life Reviews* 9: 426–453.
Wallace, R. 2013. A new formal approach to evolutionary processes in socioeconomic systems. *Journal of Evolutionary Economics* 23: 1–15.
Wallace, R. 2014a. Cognition and biology: Perspectives from information theory. *Cognitive Processing* 15: 1–12.
Wallace, R. 2014b. Embodied cognition, embodied regulation, and the data rate theorem. *Biologically Inspired Cognitive Architectures* 8: 69–80.
Wallace, R. 2015. *An Information Approach to Mitochondrial Dysfunction: Extending Swerdlow's Hypothesis*. World Scientific Series in Information Studies, vol. 4. Singapore: World Scientific.
Wallace, R., 2017, *Information Theory Models of Instabilities in Critical Systems*. World Scientific Series in Information Studies, vol. 7. Singapore: World Scientific.
Wallace, R., and M. Fullilove. 2008. *Collective Consciousness and Its Discontents*. New York: Springer.
Wallace, R., and D. Wallace. 2010. *Gene Expression and Its Discontents: The Social Production of Chronic Disease*. New York: Springer.
Wallace, R., and D. Wallace. 2013. *A Mathematical Approach to Multilevel, Multiscale Health Interventions: Pharmaceutical Industry Decline and Policy Response*. London: Imperial College Press.
Weinstein, A. 1996. Groupoids: unifying internal and external symmetry. *Notices of the American Mathematical Association* 43: 744–752.
Wilson, K. 1971. Renormalization group and critical phenomena I. Renormalization group and the Kadanoff scaling picture. *Physical Review B* 4: 3174–3183.
Wilson, A., and S. Golonka. 2013. Embodied cognition is not what you think it is. *Frontiers in Psychology*. doi: 10.3389/fpsyg.2013.00058.
Wolpert, D., and W. MacReady. 1995. No free lunch theorems for search. Santa Fe Institute, SFI-TR-02-010.
Wolpert, D., and W. MacReady. 1997. No free lunch theorems for optimization. *IEEE Transactions on Evolutionary Computation* 1: 67–82.
You, K., and L. Xie. 2013. Survey of recent progress in networked control systems. *Acta Automatica Sinica* 39: 101–117.
Yu, S., and P. Mehta. 2010. Bode-like fundamental performance limitations in control of nonlinear systems. *IEEE Transactions on Automatic Control* 55: 1390–1405.

Chapter 7
Tools for the Future: Hidden Symmetries

Summary Adopting Maturana's and Varela's perspective on the necessity of cognition—or Rosen's view of "anticipation"—at every scale and level of organization of the living state it is possible to extend Yeung's relation between information inequalities and finite group structure to a fundamental duality between an information-theoretic characterization of cognition in gene expression and the groupoid extension of simple symmetries. It appears that gene expression, responding to and hence "anticipating" environmental clues, in a large sense, may sometimes be characterized by groupoids of increasing complexity constructed from underlying finite groups. While higher organisms at later developmental stages may not be described so simply, Yeung's powerful results suggest the existence of regularities built from finite groups during certain developmental periods via an extension of the spontaneous symmetry breaking/making formalism familiar from physical theory. Essentially, the S-shaped developmental curve serves as a temperature analog driving increasing levels of deeply underlying topological symmetries. This suggests, as in the characterization of complex geometric objects by "simpler" structures in algebraic topology, that a shift of perspective from gene expression networks and their dynamics to their underlying symmetries may provide deeper insight to ontology and its dysfunctions.

> The inference of gene regulatory networks... from experimental observations is at the heart of systems biology. This includes the inference of both network topology and its dynamics. (Vera-Licona et al. 2014)
>
> ... [P]eriodic solutions... can occur in a network with no symmetry... generated by symmetry on a quotient network. (Golubitsky et al. 2012)
>
> ... [I] was astonished to find this profusion of anticipatory behavior at all levels of biological organization. (Rosen 2012)

7.1 Introduction

Recent work by Hendriks et al. (2014) on gene expression in the relatively simple reference organism *C. elegans* uncovered unexpected periodicities. The experimenters found that their genome-wide and temporally highly resolved gene

R. Wallace, *Computational Psychiatry*, DOI 10.1007/978-3-319-53910-2_7

expression studies revealed extensive periodic gene expression during *C. elegans* larval development. This observation affected a fifth of expressed genes, revealing robust, transcriptionally driven oscillations across a continuum of phases that result in periodic translation, promoting periodic developmental processes. These unanticipated results highlight dynamics and complexity of gene expression patterns during *C. elegans* development. Hendriks et al. propose that a unique combination of features makes these oscillations a powerful model to study coordinated gene expression in an animal, i.e., that insights into the mechanisms that achieve robust phase locking and broad distribution of phases may be illuminating for the understanding of coordinated gene expression more generally. They find it striking that oscillations are robustly detectable in RNA from whole animals, suggesting the presence of effective mechanisms, yet to be uncovered, that coordinate oscillations spatially and temporally.

The associated commentary by Laxman et al. (2014) explores numerous similar examples, finding that such transcriptional clocks allow organisms to temporally compartmentalize biological processes in synchrony with environmental cues. As they note, in photosynthetic cyanobacteria, the periodic expression of clusters of circadian clock-controlled genes enables the organism to generate energy and grow while temporally separating incompatible chemical processes, a mechanism also observed in plants. Such cycles, they assert, can be considered as being polytopic, regulating multiple processes through temporal compartmentalization of metabolism, suggesting that cyclic changes in a cell's metabolic state can drive such biological oscillations. Their reference list is striking.

Theoretical work on symmetries associated with periodicities and oscillations in network systems lays a foundation for understanding the yet-to-be-uncovered mechanisms. Following closely the arguments of Golubitsky and Stewart (2006), a formal theory of symmetries of networks of coupled dynamical systems, using the permutation group of the nodes that preserve the network topology, has existed for some time. Global network symmetries impose strong constraints on their associated dynamical systems, characterizing equilibria, periodic states, heteroclinic cycles, and chaotic states. The group symmetries of the network can lead to synchrony, phase relations, resonances, and synchronous or cycling chaos. However, group symmetry is, they argue, too restrictive an assumption, and network theory should be more comprehensive. They use an extension of the group-theoretic notion of symmetry to replace global symmetries with bijections between appropriate input subsets of the directed edges of the network. The symmetry group then is a *groupoid*, an algebraic structure that resembles a group but is different in that the product of two elements is not necessarily defined. Golubitsky and Stewart argue that use of groupoids makes it possible to extend group-theoretic methods to more general networks, permitting full classification of dominant patterns of synchrony in terms of the combinatorial structure of the network. Golubitsky et al. (2012) find that periodicities can be driven on a network without symmetries by those characterizing a quotient network .

Adapting an information theory treatment of cognition and Yeung's (2008) analysis of the consonance between information theory inequalities and results from the theory of finite groups, it becomes possible to expand the underlying arguments to more complex systems than *C. elegans* and yeasts.

7.2 The Cognitive Paradigm for Gene Expression

A direct approach to the dynamics of the regulation of gene expression invokes an extension of the "cognitive paradigm" of Atlan and Cohen (1998), who recognized that the immune response is not merely an automatic reflex, but involves active choice of a particular response to insult from a larger repertoire of possible responses, according to comparison with an internal learned or inherited picture of the world. Choice reduces uncertainty and implies the existence of an underlying information source (Wallace 2012a). Similar perspectives apply to gene expression (Wallace and Wallace 2010). That is, gene expression is also a cognitive phenomenon, analogous to, if different from, the immune response since, at developmental branch points, choice must be made regarding which genes to activate and which to suppress.

The Atlan/Cohen perspective is recognizably similar to Rosen's (2012, p. 313) characterization of an "anticipatory system" as one

> ...containing a predictive model of itself and/or its environment, which allows it to change state in an instant in accord with the model's predictions...

Following Chap. 1, given an information source associated with a cognitive gene expression system—called "dual" to it—an equivalence class algebra can be constructed by choosing different system origin states a_0 and defining the equivalence of two subsequent states at times $m, n > 0$, written as a_m, a_n, by the existence of high probability "meaningful" paths connecting them to the same origin point. Disjoint partition by equivalence class, analogous to orbit equivalence classes in dynamical systems, defines a symmetry groupoid associated with the cognitive process. Again, groupoids are generalizations of the group concept in which there is not necessarily a product defined for each possible element pair (Weinstein 1996).

The equivalence classes define a set of cognitive dual information sources available to the gene regulation system, creating a large groupoid, with each orbit corresponding to a transitive groupoid whose disjoint union is the full groupoid. Each subgroupoid has its own dual information source, and larger groupoids will have richer dual information sources than smaller.

Let X_{G_i} be the gene expression system's dual information source associated with the groupoid element G_i, and take Y as the information source associated with an embedding environment that, in a large sense, signals the developmental process. Chapter 1 details how environmental regularities imply the existence of an environmental information source.

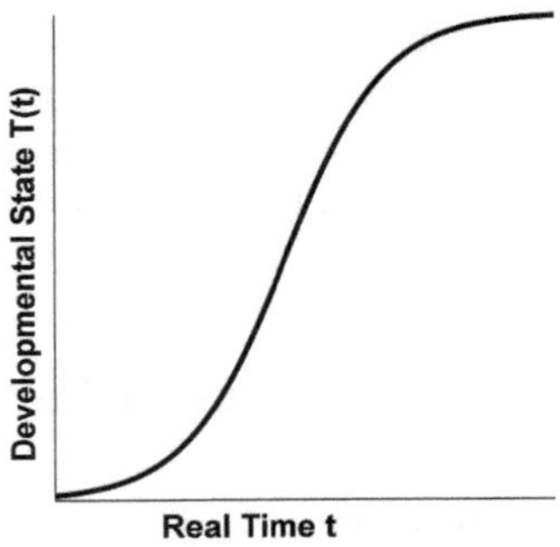

Fig. 7.1 Dimensionless "developmental state" $T(t)$ as a function of real time t

A "free energy" Morse Function (Pettini 2007) can now be constructed using a pseudoprobability argument.

Let $H(X_{G_i}, Y) \equiv H_{G_i}$ be the joint uncertainty of the two information sources. A Boltzmann-like pseudoprobability is written as

$$P[H_{G_i}] = \frac{\exp[-H_{G_i}/T(t)]}{\sum_j \exp[-H_{G_j}/T(t)]} \tag{7.1}$$

where $T(t)$ is a dimensionless "developmental state" characteristic of the organism, t is the real time, and the sum is over the different possible cognitive modes of the full system. $T(t)$ would be expected to have the usual S-shaped form, for example,

$$T(t) = \frac{\kappa_1}{1 + \kappa_2 \exp[-\alpha t]} \tag{7.2}$$

with $\alpha > 0$, $\kappa_2 \gg \kappa_1 > 0$, leading to a dynamic like that of Fig. 7.1. α has the dimension of a rate and the κ_i are numerical.

A Morse Function $\mathcal{F}$—analogous to free energy in a physical system—can be defined in terms of the "partition function" in the denominator of Eq. (7.1) as

$$\exp[-\mathcal{F}/T(t)] \equiv \sum_j \exp[-H_{G_j}/T(t)] \tag{7.3}$$

Given a groupoid structure characteristic of gene expression networks as a generalization of the simple symmetry group, it becomes possible to apply an extension of Landau's picture of phase transition in a physical system (Pettini 2007). In Landau's "spontaneous symmetry breaking," phase transitions driven by temperature changes occur as alteration of system symmetry, with higher energies at higher temperatures being more symmetric. The proposed shift between symmetries is highly punctuated in the temperature. Typically, there are only a very limited number of possible phases. Here, the developmental index plays the role of temperature. That is, increasing $T(t)$ leads to staged increase in the complexity of cognitive developmental process—as affected by signals from the embedding environment determining norms of reaction—although development itself is, of course, uniform. Some further effort, however, suggests the possibility of a more specific approach.

It is worth reiterating that incorporation of the environmental information source Y into the calculation of $\mathcal{F}$ implies that norms of reaction will be incorporated into the developmental trajectory of the organism. Different norms would be associated with different sequences of groupoids observable during development, particularly at critical ontological branch points.

7.3 The Group Structures of Information Processes

There is some theoretical support for this approach. Yeung (2008) has explored the deep relationship between information theory and the theory of finite groups, mirroring the relation between synchronicity in networks and their permutation symmetries as discussed at length in Golubitsky and Stewart (2006). For example, given two random variables X_1 and X_2 with Shannon uncertainties $H(X_1)$ and $H(X_2)$, the information theory chain rule (Cover and Thomas 2006) states that, for the joint uncertainty $H(X_1, X_2)$,

$$H(X_1) + H(X_2) \geq H(X_1, X_2) \tag{7.4}$$

Similarly, let G be any finite group, and G_1, G_2 be subgroups of G. Let $|G|$ represent the order of a group—the number of elements. Then the intersection $G_1 \cap G_2$ is also a subgroup, and a "group inequality" can be derived that is the precise analog of Eq. (7.4):

$$\log\left[\frac{|G|}{|G_1|}\right] + \log\left[\frac{|G|}{|G_2|}\right] \geq \log\left[\frac{|G|}{|G_1 \cap G_2|}\right] \tag{7.5}$$

Yeung defines a probability for a pseudorandom variate associated with a group G as $\Pr\{X = a\} = 1/|G|$. This allows construction of a group-characterized information source, noting that, in general, the joint uncertainty of a set of random variables is not necessarily the logarithm of a rational number. The surprising result Yeung (2008) ultimately establishes is a one-to-one correspondence between unconstrained information inequalities—generalizations of Eq. (7.4)—and finite group inequalities: unconstrained inequalities can be proved by techniques in group theory, and many group-theoretic inequalities can be proven by techniques of information theory. Yeung uses an obscure unconstrained information inequality to derive, in his Eq. (16.116), a complex group inequality for which, as he puts it, the "... implications in group theory are yet to be understood."

An intuitive argument in this direction actually follows from the essential mantra of algebraic topology (Hatcher 2001): forming algebraic images of topological spaces. The most basic of these images is the fundamental group, leading to Van

Kampen's theorem allowing the computation of the fundamental group of spaces that can be decomposed into simpler spaces whose fundamental group is already known. As Hatcher (2001, p. 40) puts it, "By systematic use of this theorem one can compute the fundamental groups of a very large number of spaces... [F]or every group G there is a space X_G whose fundamental group is isomorphic to G". As Golubitsky and Stewart forcefully argue, network structures and dynamics are imaged by fundamental groupoids, for which there also exists a version of the Seifert–Van Kampen theorem (Brown et al. 2011). Yeung's (2008) results suggest information theory-based "cognitive" generalizations that may include essential dynamics of gene expression and its regulation.

7.4 The Topology of "Code Networks"

Curiously parallel results emerge from Tlusty's (2007, 2008) error-minimization analysis of the genetic code, in which a "network" of codons is embedded in a larger topological structure having a fundamental group. Tlusty finds that discussion of the topology of errors portrays the codon space as a graph whose vertices are the codons. In his analysis, two codons are linked by an edge if they are likely to be confused by misreading. He assumes that two codons are most likely to be confused if all their letters except for one agree and therefore draw an edge between them. The resulting graph is, in his approach, "natural" for considering the impact of translation errors on mutations because such errors almost always involve a single letter difference, that is, a movement along an edge of the graph to a neighboring vertex.

The topology of a graph is characterized by its genus γ, the minimal number of holes required for a surface to embed the graph such that no two edges cross. The more connected that a graph is the more holes are required for its minimal embedding. The highly interconnected 64-codon graph is embedded in a hole, $\gamma = 41$ surface. The genus is somewhat reduced to $\gamma = 25$ if only 48 effective codons are considered.

Tlusty uses the extremum of an information-theoretic Morse Function to determine a single contiguous domain where a certain amino acid is encoded. Thus every mode corresponds to an amino acid and the number of modes is the number of amino acids. This compact organization is advantageous, he claims, because misreading of one codon as another codon within the same domain has no deleterious impact: if the code has two amino acids, it is the error load of an arrangement where there are two large contiguous regions, each coding for a different amino acid is much smaller than a checkerboard arrangement of the amino acids.

This result is analogous to—but significantly different from—the topological coloring problem. In the coding problem one desires maximal similarity in the colors of neighboring "countries," while the coloring problem must color neighboring countries by different colors. The number of possible amino acids in this scheme is determined by Heawood's formula (Ringel and Young 1968),

$$\mathrm{chr}(\gamma) = \mathrm{Int}\left[\frac{1}{2}(7 + \sqrt{1 + 48\gamma})\right] \tag{7.6}$$

$\mathrm{chr}(\gamma)$ is the number of "colored" regions, Int is the integer value of the enclosed expression, and γ is the genus of the surface, roughly speaking, the number of "holes." In general, $\gamma = 1 - (1/2)(V - E + F)$, where V is the number of code network vertices, E the number of network edges, and F the number of enclosed faces.

The text table below gives the first few steps in the calculation.

γ (# surface holes)	chr(γ) (# error classes)
0	4
1	7
2	8
3	9
4	10
5	11
6, 7	12
8, 9	13

Tlusty (2007) then models the emergence of the genetic code as a phase transition in a noisy information channel, using the Rate Distortion Theorem, with the optimal code is described by the minimum of a "free energy"-like functional, as above, characterizing the code's emergence as a transition akin to a phase transition in statistical physics: a supercritical phase transition is known to take place in noisy information channels. The noisy channel is controlled by a temperature-like parameter that determines the balance between the information rate and the distortion in the same way that physical temperature controls the balance between energy and entropy in a physical system. Following Tlusty's (2007) equation (2), the free energy functional has the form $D - \hat{T}S$ where D is the average error load', equivalent to average distortion in a rate distortion problem, S is the "entropy due to random drift," and $\hat{T}$ measures the strength of random drift relative to the selection force that pushes toward fitness maximization. This is essentially a Morse Function (Pettini 2007). At high $\hat{T}$, in this model, the channel is totally random and it conveys zero information. At a certain critical temperature $\hat{T}_c$ the information rate starts to increase continuously.

This is not the only such example. Hecht et al. (2004) found that protein α-helices have an inherent "code" 101100100110... where 1 indicates a polar and 0 a non-polar amino acid. Protein β-sheets have the simpler coding 10101010... Wallace (2010) extends Tlusty's topological analysis via Heawood's graph genus formula—the "magic numbers" above—to the more complicated protein folding classifications uncovered by Chou and Maggiora (1998). Going beyond the four classes of Levitt and Chothia (1976)—α helices, β sheets, $\alpha + \beta$ and α/β structures—three more minor classifications, and possibly another three subminor classes. The globular "protein folding error code network" becomes a large connected "sphere" producing the four fundamental classes, having one minor, and possibly as many as three more subminor attachment handles, in the Morse Theory sense (Pettini 2007).

Wallace (2012b) finds that similar arguments apply to the twelve monosaccharides associated with the mammalian glycan "kelp frond" production at the surface of the cell, suggesting an underlying code network having genus 6 or 7, according to the table above.

Applying a Morse Function approach to error-limiting code networks, as Wallace (2015) notes,

1. The genus of the embedding surface for a topological code can be expressed in terms of the Euler characteristic of the manifold, $\gamma = 1 - \frac{1}{2}\chi$.
2. χ can be expressed in terms of the cohomology structure of the manifold (Lee 2000, Theorem 13.38). In particular, for the code manifold, $\chi = V - E + F$ where V is the number of code network vertices, E the number of edges, and F the number of faces.
3. The fundamental group of a closed, orientable surface of genus γ is the quotient of the free group on the 2γ generators $a_1, .., a_\gamma, b_1, .., b_\gamma$ by the normal subgroup generated by the product of the commutators

$$a_1 b_1 a_1^{-1} b_1^{-1}, \ldots, a_\gamma b_\gamma a_\gamma^{-1} b_\gamma^{-1}$$

   This is a standard construction (e.g., Lee 2000).

As codes become more complex, the appropriate quotient groups must become richer, consonant with "spontaneous symmetry" arguments.

A similar approach will apply to other kinds of networks—direct biochemical or indirect nets of interacting information sources—that are associated with gene expression and its control.

In sum, the developmental state index $T(t)$ from Eqs. (7.1)–(7.3) will serve as a temperature analog driving increased levels of deep network topological symmetry.

7.5 Expanding the Model

Equation (7.4)—and its symmetry expression in Eq. (7.5)—has profound implications for development. In essence, one information source can act as an "information catalyst" to direct, or at least influence, the trajectory of another.

At some developmental time $T(t)$, a Boltzmann probability density can be constructed in terms of the rate of metabolic free energy consumed by development. For an information source X, having source uncertainty $H(X)$ and metabolic free energy (MFE) available at some rate M,

$$\dot{P}[X] \equiv \frac{\exp[-H/\kappa M]}{\int \exp[-H/\kappa M]dH} \tag{7.7}$$

Assuming the translation rate of MFE into information is very small, to first order $\hat{H} \equiv \int H\dot{P}[H]dH \approx \kappa M$, and, using Eq. (7.4),

$$\begin{aligned} \hat{H}(X,Y) &\leq \hat{H}(X) + \hat{H}(Y) \\ M_{X,Y} &\leq M_X + M_Y \end{aligned} \tag{7.8}$$

A more comprehensive argument might involve the maximum capacities for developmental channels under the Rate Distortion Theorem, in which case H in Eq. (7.7) is replaced by a Rate Distortion Function R integrated over the range $0 \to \infty$, so that $\hat{R} = \kappa M$.

As a consequence of the information inequality of Eq. (7.4)—inherent in the symmetry relation of Eq. (7.5)—allowing crosstalk between information sources consumes less MFE than isolating information sources, something that confounds electrical engineers attempting to isolate individual signals. For developing organisms, however, information catalysis becomes a central tool for gene expression and its control. It seems likely that other expressions of the relation between information inequalities and group inequalities will sometimes have similar importance.

We can now iterate the argument across developmental time $T(t)$, in terms of a network of interacting information sources rather than a network of biochemical reactions, as is the current practice. We thus, in a sense, extend the idea of a "quotient network" from Golubitsky and Stewart (2006), involving equivalence classes of network nodes. Golubtisky and Stewart demonstrate that dynamics of the quotient network impose themselves onto the original network, although there can be lower level network dynamics that are not reflected in the quotient. This is, perhaps, analogous to "unconscious" vs "conscious" processes in the brains of higher animals: consciousness is taken as a function of "global workspace" dynamics at higher levels than individual cognitive modules. Again, see Chap. 1.

For a given underlying network topology of n interacting information sources representing cognitive submodules of a developing organism at time $T(t)$, there will be some average crosstalk between them, say ϵ_n. For that given network topology—remember, a network of interacting information sources—there will be some critical value ϵ_n^C at which a giant component (GC) emerges. That is, at ϵ_n^C, a very large number of cognitive developmental processes become linked across the developing organism into a joint GC information source. For random networks this phenomenon has been well studied and the conditions under which it occurs are well understood (Wallace 2012a). While other topologies impose different detailed conditions, the punctuated emergence of a GC is almost universal across a very large class of networks. If $\epsilon > \epsilon_C$ at some $T(t)$, then individual cognitive developmental modules become coordinated across the organism. Conversely, different network topologies can impose different critical crosstalk values for onset of the GC. Topology can drive crosstalk, crosstalk can drive topology, and topologies are inherently defined by various symmetry indices, both groups and groupoids.

7.6 Developmental Canalization and Directed Homotopy

Equation (7.8) suggests the possibility of an "information catalysis" in developmental trajectory via an imposed signal from embedding regulatory information sources. The analytic tool for this is an extension—or weakening—of the fundamental group associated with homotopy loops to that of a fundamental groupoid associated with *directed homotopy*, in the sense of Goubault and Raussen (2002) and Goubault (2003). More complete discussions can be found in Grandis (2009) and Fajstrup et al. (2016).

Directed homotopy is different from simple homotopy in that inherently one-way paths from one point to another are the fundamental objects, rather than loops beginning and ending at a point-of-origin. Continuous deformation of directed paths between the same beginning and end points defines equivalence classes of dihomotopy paths constituting a groupoid. In Fig. 7.2, C represents a developmental branch point leading from an initial phenotype So to two different possible phenotypes, $S1$ and $S2$. C casts a developmental shadow, and two equivalence classes of dihomotopy paths are possible. The embedding regulatory information sources, via Eq. (7.8), define one of these phenotypes as a relative energy minimum. Thus, for gene expression and its regulation, the fundamental symmetries are likely to be those of groupoids, built up via the groupoid version of the Seifert–Van Kampen theorem and by the occurrence of repeated critical points $C_1, C_2, \ldots$ as the developmental state index $T(t)$ increases.

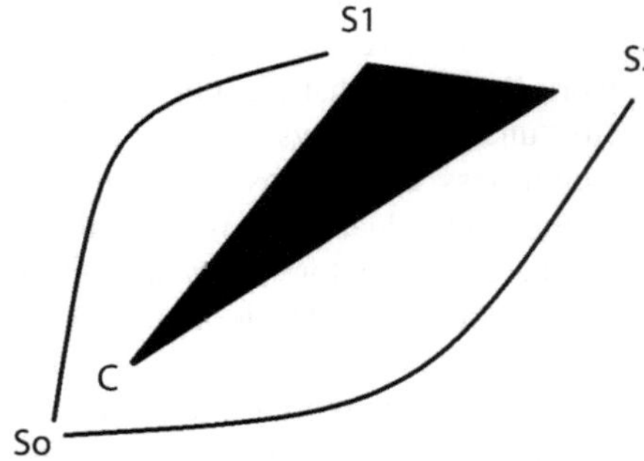

Fig. 7.2 Starting at an initial developmental phenotype So, at critical period C that casts a developmental shadow, there are two directed homotopy equivalence classes of deformable paths leading to phenotypes $S1$ and $S2$ that define a groupoid. The canalizing "choice" between them is driven by embedding regulatory information sources via the catalysis of Eq. (7.8). Repeated critical points, $C_1, C_2, \ldots$ over the developmental state $T(t)$ systematically enlarge the fundamental groupoid

7.7 Discussion

An essential inference from Yeung's results and the cognitive paradigm for gene expression is that, for many organisms, development might be associated with increasingly complex groupoid symmetries—which may sometimes be limited to simpler symmetry groups—some of which, at least, may be constructed by appropriate combinations of finite groups. In addition, it may be possible to characterize developmental trajectories by their associated groupoid sequences, keyed to different norms of reaction in which gene expression is keyed to environmental clues, in a large sense. Thus different "environmental" effects—normal or pathological—might well be associated with markedly different symmetry pathways. An interesting research question then asks if normal and pathological development have different characteristic network groupoid symmetry dynamics, and whether these can be easily identified, and perhaps modified.

An analog to this general approach can be found in the theory of error-correcting codes (as opposed to Tlusty's error-limiting codes), otherwise known as algebraic coding theory, which looks for redundancies in message coding over noisy channels enabling efficient reconstruction of lost or distorted information. Mathematical techniques involve groups, ideals, rings, algebras, and finite fields. These formalisms have generated many different codes with different capabilities and complexities, for example, BCH, Goppa, Hamming, Linear, Reed–Muller, Reed–Solomon, and so on. It may well be that the relations between groups, groupoids, and a broad spectrum of information related phenomena of interest in biology are similarly intimate.

Indeed, parallel arguments can be found in computational neuroscience. For example, Ashwin et al. (2016) infer, regarding the dynamics of neural networks, that it may be necessary to

> ...further tap into recent ideas for classifying emergent dynamics based upon the group of structural symmetries of the network...For many real-world networks, this can be decomposed into direct and wreath products of symmetric groups...

Adopting the perspective of Maturana and Varela (1980) on the necessity of cognition at every scale and level of organization of the living state, it seems possible to extend the relation between information inequalities and finite group structure to a fundamental duality between an information-theoretic characterization of cognition in gene expression and groupoids. For some, or even many, organisms, development might be characterized by groupoids of increasing complexity constructed from underlying finite groups. Higher organisms at later developmental stages may not be so simply described, but Yeung's (2008) powerful results suggest some regularities involving finite groups during certain developmental stages for some organisms.

This conjecture is a matter for empirical study, which will almost certainly find matters significantly more complicated than a simple progression to more complicated groupoids constructed from underlying finite groups: biology is inherently messy. The statistical models proposed here can, however, as with the more familiar regression models, provide benchmarks against which to compare observations on different systems under similar conditions, or the same or similar systems under different conditions.

In sum, there may be complex interactions between group and groupoid symmetries at different levels of organization and across multiple temporal and geometric scales in the living state. This may even involve the kinds of semidirect and wreath products of groups and groupoids found in the study of intrinsically disordered proteins, other nonrigid molecular structures, and symmetric networks (Wallace 2012c; MacArthur et al. 2008; Houghton 1975).

Regarding inherent symmetry, Noether's first theorem (Wikipedia 2016), as it applies in classical mechanics, states that every differentiable symmetry of the action of a physical system has a corresponding conservation law. The action of a physical system is the integral over time of a Lagrangian function, from which system behavior can be determined by the variational principle of least action. A generalization of the formulations on constants of motion in Lagrangian and Hamiltonian mechanics, the theorem does not apply to systems that cannot be modeled with a Lagrangian alone: dissipative systems with continuous symmetries need not have a corresponding conservation law, although Onsager-like linearization methods similar to those of nonequilibrium thermodynamics may sometimes be used as a first approximation (Kontogeorgaki et al. 2016).

For non-dissipative systems, rotational symmetry implies conservation of angular momentum, symmetry in time implies conservation of energy, displacement symmetry implies conservation of momentum, and so on into quantum mechanics, where matters become far more subtle. However, network and other such cognitive systems always act under dissipative circumstances, i.e., metabolic or other free energy is constantly converted to heat. Then the results of Golubitsky and Stewart and of Yeung imply—modulo the inherent dimensional collapse necessarily associated with an information source—that deep, if more complicated, symmetries may still lie behind networks of gene expression and their control.

References

Ashwin, P., S. Coombes, and R. Nicks. 2016. Mathematical frameworks for oscillatory network dynamics in neuroscience. *Journal of Mathematical Neuroscience* 6: 2. doi: 10.1186/s13408-015-0033-6.

Atlan, H., and I. Cohen. 1998. Immune information, self-organization, and meaning. *International Immunology* 10: 711–717.

Brown, R., P. Higgins, and R. Sivera. 2011. *Nonabelian Algebraic Topology: Filtered Spaces, Crossed Complexes, Cubical Homotopy Groupoids*. EMS Tracts in Mathematics, vol. 15. Zürich: European Mathematical Society. www.bangor.ac.uk/r.brown/nonab-a-t.html.

Chou, K., and G. Maggiora. 1998. Domain structural class prediction. *Protein Engineering* 11: 523–528.

Cover, T., and J. Thomas. 2006. *Elements of Information Theory*, 2nd ed. New York: Wiley.

Fajstrup, L., E. Goubault, A. Mourgues, S. Mimram, and M. Raussen. 2016. *Directed Algebraic Topology and Concurrency*. New York: Springer.

Golubitsky, M., and I. Stewart. 2006. Nonlinear dynamics of networks: The groupoid formalism. *Bulletin of the American Mathematical Society* 43: 305–364.

Golubitsky, M., D. Romano, and Y. Wang. 2012. Network periodic solutions: Patterns of phase-shift synchrony. *Nonlinearity* 25: 1045–1074.

Goubault, E. 2003. Some geometric perspectives on concurrency theory. *Homology, Homotopy, and Applications* 5: 95–136.

Goubault, E., and M. Raussen. 2002. Dihomotopy as a tool in state space analysis. In *Lecture Notes in Computer Science*, vol. 2286, 16–37. New York: Springer.

Grandis, M. 2009. *Directed Algebraic Topology: Models of Non-reversible Worlds*. New York: Cambridge University Press.

Hatcher, A. 2001. *Algebraic Topology*. New York: Cambridge University Press.

Hecht, M., A. Das, A. Go, L. Aradely, and Y. Wei. 2004. De novo proteins from designated combinatorial libraries. *Protein Science* 13: 1711–1723.

Hendriks, G., D. Gaidatzis, F. Aeschimann, and H. Grosshans. 2014. Extensive oscillatory gene expression during *C. elegans* larval development. *Molecular Cell* 53: 380–392.

Houghton, C. 1975. Wreath products of groupoids. *Journal of the London Mathematical Society* 10: 179–188.

Kontogeorgaki, S., R. Sanchez-Garcia, R. Ewing, K. Zygalakis, and B. MacArthur. 2016. Noise-processing by signalling networks. bioRxiv preprint http://dx.doi.org/10.1101/075366.

Laxman, S., B. Tu, and S. McKnight. 2014. Concerted effort: Oscillations in global gene expression during nematode development. *Molecular Cell* 53: 363–364.

Lee, J. 2000. *Introduction to Topological Manifolds*. Graduate Texts in Mathematics. New York: Springer.

Levitt, M., and C. Chothia. 1976. Structural patterns in globular proteins. *Nature* 261: 552–557.

MacArthur, B., R. Sanchez-Garcia, and J. Anderson. 2008. Symmetry in complex networks. *Discrete and Applied Mathematics* 156: 3525–3531.

Maturana, H., and F. Varela. 1980. *Autopoiesis and Cognition*. Dordrecht: Riedel.

Pettini, M. 2007. *Geometry and Topology in Hamiltonian Dynamics*. New York: Springer.

Ringel, G., and J. Young. 1968. Solutions of the Heawood map-coloring problem. *PNAS* 60: 438–445.

Rosen, R. 2012. *Anticipatory Systems: Philosophical, Mathematical, and Methodological Foundations*, 2nd ed. New York: Springer.

Tlusty, T. 2007. A model for the emergence of the genetic code as a transition in a noisy information channel. *Journal of Theoretical Biology* 249: 331–342.

Tlusty, T. 2008. A simple model for the evolution of molecular codes driven by the interplay of accuracy, diversity and cost. *Physical Biology* 5: 016001.

Vera-Licona, P., A. Jarrah, L. Garcia-Puente, J. McGee, and R. Laubenbacher. 2014. An algebra-based method of inferring gene regulatory networks. *BMC Systems Biology* 8: 37.

Wallace, R. 2010. A scientific open season. *Physics of Life Reviews* 7: 377–378.

Wallace, R. 2012a. Consciousness, crosstalk, and the mereological fallacy: An evolutionary perspective. *Physics of Life Reviews* 9: 426–453.

Wallace, R. 2012b. Extending Tlusty's rate distortion index theorem method to the glycome: Do even 'low level' biochemical phenomena require sophisticated cognitive paradigms? *BioSystems* 107: 145–152.

Wallace, R. 2012c. Spontaneous symmetry breaking in a non-rigid molecule approach to intrinsically disordered proteins. *Molecular Biosystems* 8: 374–377.

Wallace, R. 2015. Metabolic free energy and biological codes: A 'data rate theorem' aging model. *Bulletin of Mathematical Biology* 77: 879–903.

Wallace R., and D. Wallace. 2010. *Gene Expression and its Discontents: The Social Production of Chronic Disease*. New York: Springer.

Weinstein, A. 1996. Groupoids: Unifying internal and external symmetry. *Notices of the American Mathematical Association* 43: 744–752.

Wikipedia. 2016. https://en.wikipedia.org/wiki/Noether's_theorem.

Yeung, H. 2008. *Information Theory and Network Coding*. New York: Springer.

Chapter 8
Psychopathologies of Automata I: Autonomous Vehicle Systems

Summary We apply the perspectives of computational psychiatry to autonomous ground vehicles under intelligent traffic control in which swarms of interacting, self-driving devices are inherently unstable as a consequence of the basic irregularities of traffic interactions and the road network. It appears that such systems will experience large-scale failures analogous to the vast propagating fronts of power network blackouts, and possibly less benign, but more subtle patterns of pathology and consequent failure at various scales.

8.1 Introduction

Current marketing hype surrounding autonomous vehicles runs something like this:

> Since more than 90% of highway deaths are related to driver error, automating out the driver will reduce loss of life by more than 90%.

Individual vehicles, however, are nested and enmeshed within larger milieus, creating a multi-scale, multi-level synergism determining crash and fatality rates. Individual vehicles are only one part of that system, not the system itself. Asserting that part of a thing is the whole thing is the infamous *mereological fallacy*, an important tool for the construction of political lies and other forms of advertising.

One is reminded of another—if different—logical fallacy:

> If a woman can gestate a child in nine months, nine women should be able to do it in a month.

Given the inherently complicated nature of transport system safety, assertions regarding the effects of autonomous vehicles on traffic fatalities are entirely speculative and cannot be used as a sound basis for policy development.

Here, we ask a more fundamental question: are large-scale autonomous vehicle systems actually practical, particularly in the context of a rapidly deteriorating social and physical infrastructure? To do this, we examine an "end stage" limit in which many different kinds of vehicles communicate with each other (V2V), and with an intelligent roadway infrastructure (V2I), both embedded in a highly stochastic environment.

R. Wallace, *Computational Psychiatry*, DOI 10.1007/978-3-319-53910-2_8

In effect, we adapt mathematical models from computational psychiatry to explore the dynamics of a rapid-acting, inherently unstable command, communication and control system (C^3) that is cognitive in the sense that it must, in an appropriate "real time," evaluate a large number of possible actions and choose a small subset for implementation. Such choice decreases uncertainty, in a formal manner, and reduction in uncertainty implies the existence of an information source (Wallace 2012, 2015a).

We particularly study autonomous V2V/V2I systems through the prism of the Data Rate Theorem (Nair et al. 2007), extending the argument to more general phase transition analogs, and developing statistical tools useful at different scales and levels of organization.

8.2 Central Problems

V2V/V2I autonomous systems operate along geodesics in a densely convoluted "map quotient space" that is in contrast to the much more straightforward problem of air traffic control, where locally stable vehicle paths are seen as thick braid geodesics in a simpler Euclidean quotient space (Hu et al. 2001). Such geodesics are generalizations of the streamline characteristics of hydrodynamic flow (Landau and Lifshitz 1987).

Hu et al. (2001) show that, in the context of air traffic control, finding collision-free maneuvers for multiple agents on a Euclidean plane surface $\mathcal{R}^2$ is the same as finding the shortest geodesic in a particular manifold with nonsmooth boundary. Given n vehicles, the Hu geodesic is calculated for the topological quotient space $\mathcal{R}^{2n}/W(r)$, where $W(r)$ is defined by the requirement that no vehicles are closer together than some critical Euclidean distance r. For autonomous ground vehicles, $\mathcal{R}^2$ must be replaced by a far more topologically complex roadmap space $\mathcal{M}^2$ subject to traffic jams and other "snowflake" condensation geometries in real time. Geodesics for n vehicles are then in a highly irregular quotient space $\mathcal{M}^{2n}/W(r)$ whose dynamics are subject to phase transitions in vehicle density ρ (Kerner and Klenov 2009; Kerner et al. 2015; Jin et al. 2013) that, we will show, represent cognitive groupoid symmetry breaking.

In first order, given the factoring out of most of the topological structure by the construction of geodesics in the quotient space $\mathcal{M}^{2n}/W(r)$, the only independent system parameter is the density of vehicles per unit length, which we call ρ. Figure 8.1 shows, for streets in Rome, Japan, and Flanders, the number of vehicles per unit time as a function of, respectively, vehicles per mile, per kilometer, and percent occupancy: the "fundamental diagram" of traffic flow. There is a clear "phase transition" at about 40 vehicles/mile for the former two examples, and at about 10% occupancy for the latter.

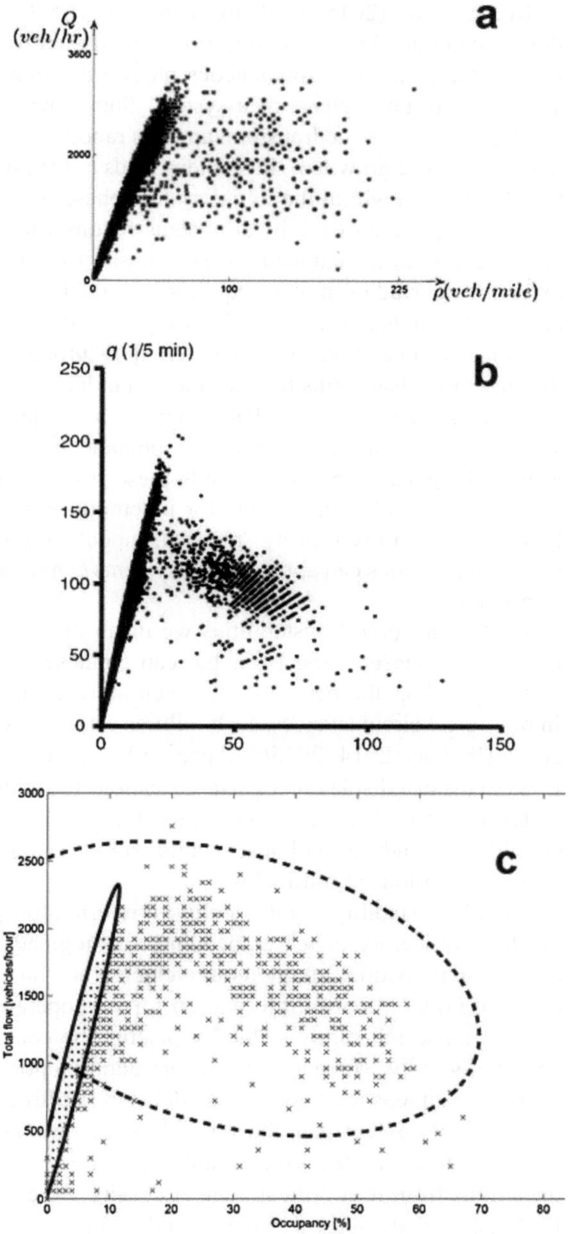

Fig. 8.1 (**a**) Vehicles per hour as a function of vehicle density per mile for a street in Rome (Blandin et al. 2011). Both streamline geodesic flow and the phase transition to "crystallized" turbulent flow at critical traffic density are evident at about 40 vehicle/mile. Some of the states may be "supercooled," i.e., delayed "crystallization" in spite of high traffic density. "Fine structure" can be expected within both geodesic and turbulent modes. (**b**) One month of data at a single point on a Japanese freeway, flow per 5 min vs. vehicles per kilometer. The critical value is about 25 vehicles/km = 39.1 vehicles/mile (Sugiyama et al. 2008). (**c**) 49 Mondays on a Flanders freeway. The ellipses contain 97.5% of data points for the free flow and congested regimes (Maerivoet and De Moor 2006). Breakdown begins just shy of 10% occupancy

We shall extend a simple vehicle density measure to a more complicated unsymmetric density matrix that includes multimodal vehicle indices, an inverse measure of roadway quality, and can be extended to measures of information channel congestion.

Kerner et al. (2015) explicitly apply insights from statistical physics to traffic flow, finding that in many equilibrium and dissipative metastable systems of natural science there can be a spontaneous phase transition from one metastable phase to another metastable phase of a system. Such spontaneous phase transition occurs when a nucleus for the transition appears randomly in an initial metastable phase of the system: The growth of the nucleus leads to the phase transition. The nucleus can be a fluctuation within the initial system phase whose amplitude is equal or larger than an amplitude of a critical nucleus required for spontaneous phase transition. Nuclei for such spontaneous phase transitions can be observed in empirical and experimental studies of many equilibrium and dissipative metastable systems. There can also be another source for the occurrence of a nucleus, rather than fluctuations: A nucleus can be induced by an external disturbance applied to the initial phase. In this case, the phase transition is called an induced phase transition.

A Data Rate Theorem (DRT) approach to stability and flow of autonomous vehicle/traffic control systems, via spontaneous symmetry breaking in cognitive groupoids, generalizes and extends these insights, implying a far more complex picture of control requirements for inherently unstable systems than is suggested by the Theorem itself, or by "physics" models of phase transition. That is, "higher order" instabilities can appear. Such systems can require inordinate levels of control information.

By "higher order" instabilities we mean that C^3 systems may remain "stable" in the strict sense of the DRT, but can collapse into a ground state analogous to certain psychopathologies, or into even more complicated pathological dynamics. In biological circumstances, such failures can be associated with the onset of senescence (Wallace 2014, 2015b). Apparently, rapidly responding inherently unstable, C^3 systems can display recognizable analogs to senility under fog-of-war demands.

Using these ideas, it becomes possible to formally represent the interaction of cognitive ground state collapse in autonomous vehicle/intelligent road systems with critical transitions in traffic flow.

Defining "stability" as the ability to return, after perturbation, to the streamline geodesic trajectory of the embedding, topologically complex, road network, it is clear that individual autonomous vehicles are inherently unstable and require a constant flow of control information for safe operation, unlike aircraft that can, in fact, be made inherently stable by placing the center of pressure well behind the center of gravity. There is no such configuration possible for ground-based vehicles following sinuous road geometries in heavy, shifting, traffic.

Recall Fig. 8.1. Again, the vertical axis shows the number of vehicles per hour, the horizontal, the density of vehicles per mile. The streamline geodesic flow and deviations from it at critical vehicle density are evident. Some of the phases may be "supercooled"—fast-flowing "liquid" at higher-than-critical densities. Additional "fine structure" should be expected within both geodesic and turbulent modes.

Classic traffic flow models based on extensions of hydrodynamic perspectives involving hyperbolic partial differential equations (HPDEs) can be analogously factored using the methods of characteristic curves and Riemann invariants—streamlines (Landau and Lifshitz 1987). Along characteristic curves, HPDEs are projected down to ordinary differential equations (ODEs) that are usually far easier to solve. The ODE solution or solutions can then be projected upward as solutions to the HPDEs. Here, we will show, reduction involves expressing complex dynamics in terms of relatively simple stochastic differential equations and their stability properties. Those stability properties, marking the onset of "turbulence," will be of central interest.

Taking a somewhat larger view, cognitive phase transitions in V2V/V2I systems, particularly ground state collapse to some equivalent of "all possible targets are enemies," should become synergistic with more familiar traffic flow phase transitions to produce truly monumental traffic jams.

A heuristic argument is as follows: Consider a random network of roads between nodal points—intersections. If the average probability of passage falls below a critical value, the Erdos/Renyi "giant component" that connects across the full network breaks into a set of disjoint connected equivalence class subcomponents, with "bottlenecks" at which traffic jams occur marking corridors between them. Li et al. (2015), in fact, explicitly apply a percolation model to explain this effect for road congestion in a district of Beijing. The underlying road network is shown in Fig. 8.2, and in Fig. 8.3 a cross section taken during rush hour showing disjoint sections when regions with average velocity below 40% of observed maximum for the road link have been removed.

Such equivalence classes define a groupoid, an extension of the idea of a symmetry group (Weinstein 1996). Below, we will define the cognitive groupoid to be associated with a C^3 structure, here a system of autonomous vehicles linked together in a V2V "swarm intelligence" embedded in a larger vehicle to infrastructure (V2I) traffic management system. Individual vehicle spacings, speed, acceleration, lane-change, and so on are determined by this encompassing distributed cognitive machine that attempts to optimize traffic flow and safety.

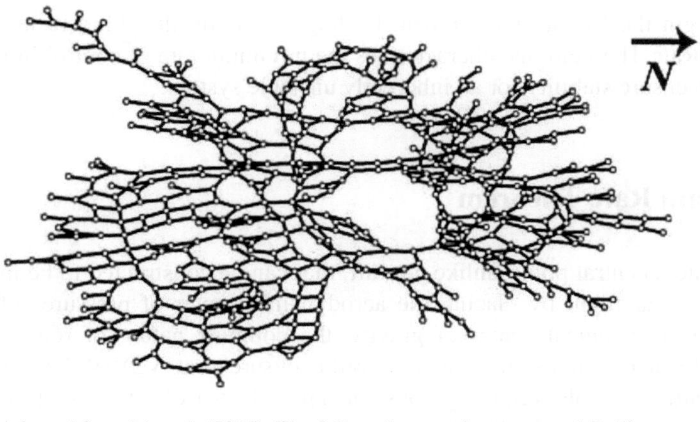

Fig. 8.2 Adapted from Li et al. (2015). The full road network near central Beijing

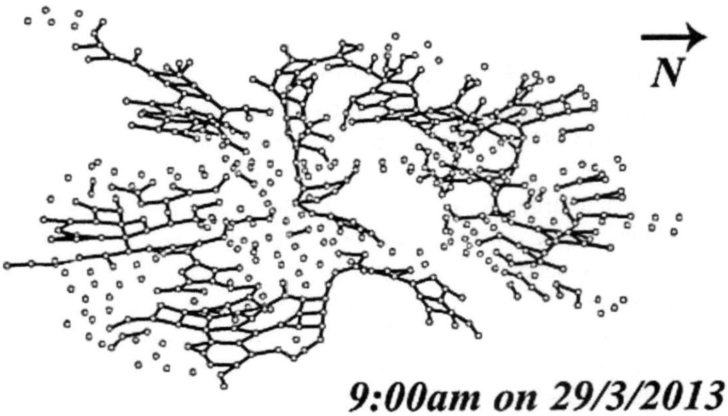

Fig. 8.3 Adapted from Li et al. (2015). Disconnected subcomponents of the Beijing central road network at rush hour. Sections with average vehicle velocity less than 40% of maximum observed have been removed. Disjoint pieces form equivalence classes that permit definition of a groupoid symmetry

The associated individual groupoids are the basic transitive groupoids that build a larger composite groupoid of the cognitive system (Wallace 2012, 2015a). Thus, under declining probability of passage, related to traffic congestion and viewed as a temperature analog, this "vehicle/road" groupoid undergoes a symmetry breaking transition into a combined cognitive ground state collapse and traffic jam mode—essentially a transition from "laminar" geodesic to "turbulent" or "crystallized" flow. Autonomous vehicle systems that become senile under fog-of-war demands will likely trigger traffic jams that are far different from those associated with human-controlled vehicles. There is no reason to believe that such differences will be benign.

We begin the formal development leading to this result with a restatement of the Data Rate Theorem that characterizes the minimum rate of control information needed to ensure stability for an inherently unstable system.

8.3 Data Rate Theorem

To reiterate a central point, unlike aircraft, that can be constructed to be inherently stable in linear flight by placing the aerodynamic center of pressure sufficiently behind the mechanical center of gravity, the complex nature of road geometry and the local dynamics of vehicular traffic ensure that V2V/V2I systems will be inherently unstable, requiring constant input of control information to prevent crashes, traffic jams, and other tie-ups.

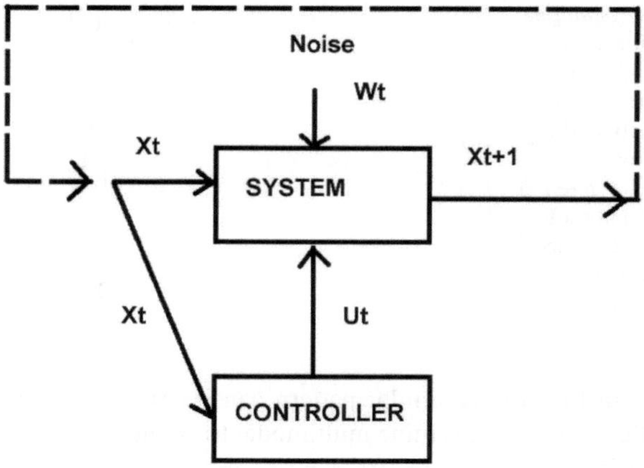

Fig. 8.4 A linear expansion near a nonequilibrium steady state of an inherently unstable control system, for which $x_{t+1} = \mathbf{A}x_t + \mathbf{B}u_t + W_t$. $\mathbf{A}, \mathbf{B}$ are square matrices, x_t the vector of system parameters at time t, u_t the control vector at time t, and W_t a white noise vector. The Data Rate Theorem states that the minimum rate at which control information must be provided for system stability is $\mathcal{H} > \log[|\det[\mathbf{A}^m|]$, where $\mathbf{A}^m$ is the subcomponent of $\mathbf{A}$ having eigenvalues ≥ 1

The Data Rate Theorem (Nair et al. 2007) establishes the minimum rate at which externally supplied control information must be provided for an inherently unstable system to maintain stability. Given the linear expansion near a nonequilibrium steady state, an n-dimensional vector of system parameters at time t, x_t, determines the state at time $t + 1$ according to the model of Fig. 8.4, so that

$$x_{t+1} = \mathbf{A}x_t + \mathbf{B}u_t + W_t \tag{8.1}$$

where $\mathbf{A}, \mathbf{B}$ are fixed $n \times n$ matrices, u_t is the vector of control information, and W_t is an n-dimensional vector of white noise. The Data Rate Theorem (DRT) under such conditions states that the minimum control information rate $\mathcal{H}$ is determined by the relation

$$\mathcal{H} > \log[|\det(\mathbf{A}^m)|] \equiv a_0 \tag{8.2}$$

where, for $m \leq n$, $\mathbf{A}^m$ is the subcomponent of $\mathbf{A}$ having eigenvalues ≥ 1. The right-hand side of Eq. (8.2) is interpreted as the rate at which the system generates "topological information." The proof of Eq. (8.2) is not particularly straightforward (Nair et al. 2007), and the Mathematical Appendix uses the Rate Distortion Theorem (RDT) to derive a more general version of the DRT.

For a simple traffic flow system on a fixed highway network, the source of "topological information" is the linear vehicle density ρ. The "fundamental diagram" of traffic flow studies relates the total vehicle flow to the linear vehicle

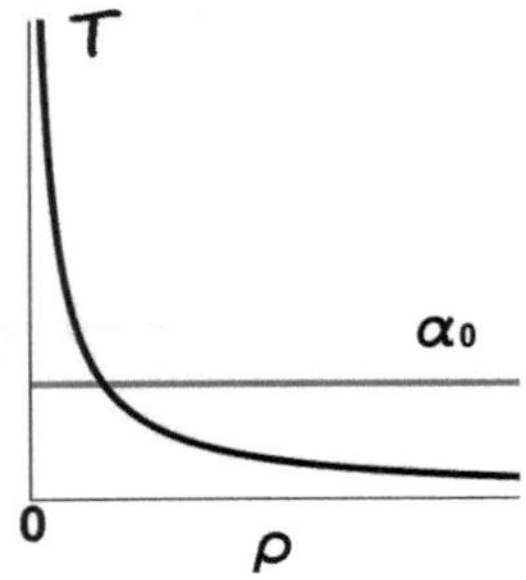

Fig. 8.5 The *horizontal line* represents the critical limit a_0. If $\kappa_2/\kappa_4 \gg \kappa_1/\kappa_3$, at some intermediate value of linear traffic density ρ, the temperature analog $\mathcal{T} \equiv (\kappa_1\rho + \kappa_2)/(\kappa_3\rho + \kappa_4)$ falls below that limit, traffic flow becomes "supercooled," and traffic jams become increasingly probable

density, shown in Fig. 8.1. A similar pattern can be expected from "macroscopic fundamental diagrams" that examine multimodal travel networks (Geroliminis et al. 2014; Chiabaut 2015).

Given ρ as the critical traffic density parameter, we can extend Eq. (8.2) as

$$\mathcal{H}(\rho) > f(\rho)a_0 \tag{8.3}$$

where a_0 is a road network constant and $f(\rho)$ is a positive, monotonically increasing function. The Mathematical Appendix uses a Black–Scholes model to approximate the "cost" of $\mathcal{H}$ as a function of the "investment" ρ. The first approximation is linear, so that $\mathcal{H} \approx \kappa_1\rho + \kappa_2$. Expanding $f(\rho)$ to similar order,

$$f(\rho) \approx \kappa_3\rho + \kappa_4 \tag{8.4}$$

the limit condition for stability becomes

$$\mathcal{T} \equiv \frac{\kappa_1\rho + \kappa_2}{\kappa_3\rho + \kappa_4} > a_0 \tag{8.5}$$

For $\rho = 0$, the stability condition is $\kappa_2/\kappa_4 > a_0$. At large ρ this becomes $\kappa_1/\kappa_3 > a_0$. If $\kappa_2/\kappa_4 \gg \kappa_1/\kappa_3$, the stability condition may be violated at high traffic densities, and instability becomes manifest, as at the higher ranges of Fig. 8.1. See Fig. 8.5.

8.4 Multimodal Traffic on Bad Roads

For vehicles embedded in a larger traffic stream there are many other possible critical densities that must interact: different kinds of vehicles per linear mile, V2V/V2I communications bandwidth crowding, and an inverse index of roadway quality that one might call "potholes per mile," and so on. There is not, then, a simple "density" index, but rather a possibly large non-symmetric density matrix $\hat{\rho}$ having interacting components with $\rho_{i,j} \neq \rho_{j,i}$.

Can there still be some scalar "ρ" under such complex circumstances so that the conditions of Fig. 8.5 apply? An $n \times n$ matrix $\hat{\rho}$ has n invariants $r_i, i = 1..n$, that remain fixed when "principal component analysis" transformations are applied to data, and these can be used to construct an invariant scalar measure, using the polynomial relation

$$p(\lambda) = \det(\hat{\rho} - \lambda I) = \lambda^n + r_1\lambda^{n-1} + \cdots + r_{n-1}\lambda + r_n \quad (8.6)$$

det is the determinant, λ is a parameter, and I the $n \times n$ identity matrix. The invariants are the coefficients of λ in $p(\lambda)$, normalized so that the coefficient of λ^n is 1. Typically, the first invariant will be the matrix trace and the last $\pm$ the matrix determinant.

For an $n \times n$ matrix it then becomes possible to define a composite scalar index Γ as a monotonic increasing function of these invariants

$$\Gamma = f(r_1, \ldots, r_n) \quad (8.7)$$

The simplest example, for a 2×2 matrix, would be

$$\Gamma = m_1\mathrm{Tr}[\hat{\rho}] + m_2|\det[\hat{\rho}]| + m_3\mathrm{Tr}[\hat{\rho}]|\det[\hat{\rho}]| \quad (8.8)$$

for positive m_i. Recall that, for $n = 2$, $\mathrm{Tr}[\hat{\rho}] = \rho_{11} + \rho_{22}$ and $\det[\hat{\rho}] = \rho_{11}\rho_{22} - \rho_{12}\rho_{21}$. In terms of the two possible eigenvalues α_1, α_2, $\mathrm{Tr}[\hat{\rho}] = \alpha_1 + \alpha_2$, $\det[\hat{\rho}] = \alpha_1\alpha_2$.

Again, an $n \times n$ matrix will have n such invariants from which a scalar index Γ can be constructed.

In Eq. (8.5) defining $\mathcal{T}$, ρ is then replaced by the composite density index Γ.

The method is a variant of the "Rate Distortion Manifold" of Glazebrook and Wallace (2009) or the "Generalized Retina" of Wallace and Wallace (2013, Sect. 10.1) in which high dimensional data flows can be projected down onto lower dimensional, shifting, tunable "tangent spaces" with minimal loss of essential information.

8.5 The Dynamics of Service Collapse

We next examine the dynamics of $\mathcal{T}(\Gamma)$ itself under stochastic circumstances. We begin by asking how a control signal u_t in Fig. 8.4 is expressed in the system response x_{t+1}. We suppose it possible to deterministically retranslate an observed sequence of system outputs $X^i = x_1^i, x_2^i, \ldots$ into a sequence of possible control signals $\hat{U}^i = \hat{u}_0^i, \hat{u}_1^i, \ldots$ and to compare that sequence with the original control sequence $U^i = u_0^i, u_1^i, \ldots$, with the difference between them having a particular value under some chosen distortion measure and hence having an average distortion

$$\langle d \rangle = \sum_i p(U^i)d(U^i, \hat{U}^i) \quad (8.9)$$

where $p(U^i)$ is the probability of the sequence U^i and $d(U^i, \hat{U}^i)$ is the distortion between U^i and the sequence of control signals that has been deterministically reconstructed from the system output.

We can then apply a classic Rate Distortion argument. According to the Rate Distortion Theorem, there exists a Rate Distortion Function, $R(D)$, that determines the minimum channel capacity necessary to keep the average distortion below some fixed limit D (Cover and Thomas 2006). Based on Feynman's (2000) interpretation of information as a form of free energy, it becomes possible to construct a Boltzmann-like pseudoprobability density in the "temperature" $\mathcal{T}$ as

$$dP(R, \mathcal{T}) = \frac{\exp[-R/\mathcal{T}]dR}{\int_0^\infty \exp[-R/\mathcal{T}]dR} \tag{8.10}$$

since higher $\mathcal{T}$ must necessarily be associated with greater channel capacity.

The denominator can be interpreted as a statistical mechanical partition function, and it becomes possible to define a "free energy" Morse Function (Pettini 2007) $\mathcal{F}$ as

$$\exp[-\mathcal{F}/\mathcal{T}] \equiv \int_0^\infty \exp[-R/\mathcal{T}]dR = \mathcal{T} \tag{8.11}$$

so that $\mathcal{F}(\mathcal{T}) = -\mathcal{T}\log[\mathcal{T}]$.

Then an "entropy" can also be defined as the Legendre transform of $\mathcal{F}$,

$$\mathcal{S} \equiv \mathcal{F}(\mathcal{T}) - \mathcal{T}d\mathcal{F}/d\mathcal{T} = \mathcal{T} \tag{8.12}$$

The Onsager treatment of nonequilibrium thermodynamics (de Groot and Mazur 1984) can now be invoked, based on the gradient of $\mathcal{S}$ in $\mathcal{T}$, so that a stochastic Onsager equation can be written as

$$d\mathcal{T}_t = (\mu d\mathcal{S}/d\mathcal{T})dt + \beta\mathcal{T}_t dW_t = \mu dt + \beta\mathcal{T}_t dW_t \tag{8.13}$$

where μ is a diffusion coefficient and β is the magnitude of the impinging white noise dW_t.

Again, applying the Ito chain rule to $\log(\mathcal{T})$ in Eq. (8.13) (Protter 1990), Jensen's inequality for a concave function gives the nonequilibrium steady state (nss) expectation of $\mathcal{T}$ as

$$E(\mathcal{T}_t) \geq \frac{\mu}{\beta^2/2} \tag{8.14}$$

In the V2V/V2I context, μ is a "diffusion coefficient" representing attempts by the system to meet service demand, and β the magnitude of a traffic flow/roadway state "white noise" dW_t contrary to those attempts.

Recall that, in the multimodal extension of the model, the condition for stability is

$$\mathcal{T} \approx \frac{\kappa_1 \Gamma + \kappa_2}{\kappa_3 \Gamma + \kappa_4} > a_0$$

The inference is that sufficient system noise, β, can drive $\mathcal{T}$ below critical values in Fig. 8.5, triggering a system collapse analogous to a large, propagating traffic jam. Under real-world conditions, adequate service will simultaneously raise μ and lower β. Nonetheless, Eq. (8.14) is an expectation, *and there will always be some probability that* $\mathcal{T} < a_0$, i.e., that the condition for stability is violated. The system then becomes "supercooled" and subject to a raised likelihood of sudden, rapidly propagating, traffic jam-like condensations in the sense of Kerner et al. (2015).

8.6 Multiple Phases of Dysfunction

The DRT argument implies a raised probability of a transition between stable and unstable behavior if the temperature analog $\mathcal{T}(\Gamma)$ falls below a critical value. Kerner et al. (2015), however, argue that traffic flow can be subject to more than two phases. We can recover something similar via a "cognitive paradigm" like that used by Atlan and Cohen (1998) in their study of the immune system. They view a system as cognitive if it must compare incoming signals with a learned or inherited picture of the world, then actively choose a response from a larger set of those possible to it. V2V/V2I systems are clearly cognitive in that sense. Such choice, however, implies the existence of an information source, since it reduces uncertainty in a formal way. See Chap. 1 or Wallace (2015a,b) for details of the argument.

Given the "dual" information source associated with the inherently unstable cognitive V2V/V2I system, an equivalence class algebra can be constructed by choosing different system origin states and defining the equivalence of subsequent states at a later time by the existence of a high probability path connecting them to the same origin state. Disjoint partition by equivalence class, analogous to orbit equivalence classes in dynamical systems, defines a symmetry groupoid associated with the cognitive process. Groupoids are generalizations of group symmetries in which there is not necessarily a product defined for each possible element pair (Weinstein 1996), for example, in the disjoint union of different groups.

The equivalence classes across possible origin states define a set of information sources dual to different cognitive states available to the inherently unstable V2V/V2I system. These create a large groupoid, with each orbit corresponding to a transitive groupoid whose disjoint union is the full groupoid. Each subgroupoid is associated with its own dual information source, and larger groupoids must have richer dual information sources than smaller.

Let X_{G_i} be the system's dual information source associated with groupoid element G_i. Given the argument leading to Eqs. (8.5)–(8.7), we construct another Morse Function (Pettini 2007) as follows.

Let $H(X_{G_i}) \equiv H_{G_i}$ be the Shannon uncertainty of the information source associated with the groupoid element G_i. We define another pseudoprobability as

$$P[H_{G_i}] \equiv \frac{\exp[-H_{G_i}/\mathcal{T}]}{\sum_j \exp[-H_{G_j}/\mathcal{T}]} \tag{8.15}$$

where the sum is over the different possible cognitive modes of the full system.

Another, more complicated, "free energy" Morse Function F can then be defined as

$$\exp[-F/\mathcal{T}] \equiv \sum_j \exp[-H_{G_j}/\mathcal{T}] \tag{8.16}$$

or, more explicitly,

$$F = -\mathcal{T} \log\left[\sum_j \exp[-H_{G_j}/\mathcal{T}]\right] \tag{8.17}$$

As a consequence of the groupoid structures associated with complicated cognition, as opposed to a "simple" stable–unstable control system, we can now apply an extension of Landau's version of phase transition (Pettini 2007). Landau saw spontaneous symmetry breaking as representing phase change in physical systems, with the higher energies available at higher temperatures being more symmetric. The shift between symmetries is highly punctuated in the temperature index, here the "temperature" analog of Eq. (8.5), in terms of the scalar construct Γ, but in the context of groupoid rather than group symmetries. Usually, for physical systems, there are only a few phases possible. Kerner et al. (2015) recognize three phases in ordinary traffic flow, but V2V/V2I systems may have relatively complex stages of dysfunction, with highly punctuated transitions between them as various density indices change and interact.

Section 5.9 above examined sufficient conditions for a pathological ground state to "lock-in" and become highly resistant to managerial intervention, that is, in this context, for a highly persistent large-scale traffic jam.

In this context, Birkoff's (1960, p. 146) perspective on the central role of groups in fluid mechanics is of considerable interest:

> [Group symmetry] underlies the entire theories of dimensional analysis and modeling. In the form of "inspectional analysis" it greatly generalizes these theories... [R]ecognition of groups... often makes possible reductions in the number of independent variables involved in partial differential equations... [E]ven after the number of independent variables is reduced to one... the resulting system of ordinary differential equations can often be integrated most easily by the use of group-theoretic considerations.

We argue here that, for "cognitive fluids" like vehicle traffic flows, groupoid generalizations of group theory become central.

Decline in the richness of control information, or in the ability of that information to influence the system as measured by the "temperature" index $\mathcal{T}(\Gamma)$, can lead to punctuated decline in the complexity of cognitive process possible within the C^3 system, driving it into a ground state collapse that may not be actual "instability" but rather a kind of dead zone in which, using the armed drone example, "all possible targets are enemies." This condition represents a dysfunctionally simple cognitive groupoid structure roughly akin to certain individual human psychopathologies, as described in the previous chapters.

It appears that, for large-scale autonomous vehicle/intelligent infrastructure systems, the ground state dead zone involves massive, propagating tie-ups that far more resemble power network blackouts than traditional traffic jams. Again, the essential feature is the role of composite system "temperature" $\mathcal{T}(\Gamma)$. Most of the topology of the inherently unstable vehicles/roads system will be "factored out" via the construction of geodesics in a topological quotient space, so that $\mathcal{T}(\Gamma)$ inversely indexes the rate of topological information generation for an extended DRT.

Lowering the "temperature" $\mathcal{T}$ forces the system to pass from high symmetry "free flow" to different forms of "crystalline" structure—broken symmetries representing platoons, shock fronts, traffic jams, and more complicated system-wide patterns of breakdown.

In the next section, the underlying dynamic is treated in finer detail by viewing the initial phase transition as the first order onset of a kind of "turbulence," a transition from free flow to "flock" structures like those studied in "active matter" physics. Indeed, the traffic engineering perspective is quite precisely the inverse of mainstream active matter studies, which Ramaswamy (2010) describes as follows:

> It is natural for a condensed matter physicist to regard a coherently moving flock of birds, beasts, or bacteria as an orientationally ordered phase of living matter. ... [M]odels showed a nonequilibrium phase transition from a disordered state to a flock with long-range order... in the particle velocities as the noise strength was decreased or the concentration of particles was raised.

In traffic engineering, the appearance of such "long-range order" is the first stage of a traffic jam (Kerner and Klenov 2009; Kerner et al. 2015), a relation made explicit by Helbing (2001, Sect. VI) in his comprehensive review of traffic and related self-driven many-particle systems.

While flocking and schooling have obvious survival value against predation for animals in three-dimensional venues, long-range order—aggregation—among blood cells flowing along arteries is a blood clot and can be rapidly fatal.

8.7 Turbulence

The "free energy" function F in Eq. (8.17) can be used to explore dynamics within a particular system phase defined by the associated groupoid.

Given a vector of system parameters $\mathbf{K}$, in standard manner it is possible to define an "entropy" from F as the Legendre transform

$$S \equiv F(\mathbf{K}) - \mathbf{K} \cdot \nabla_{\mathbf{K}} F \tag{8.18}$$

and a nonequilibrium Onsager stochastic differential equation for dynamics in terms of the gradients in S (de Groot and Mazur 1984), which can be written in one dimension as

$$dK_t = [\mu \partial S/\partial K_t]dt + \sigma K_t dB_t \tag{8.19}$$

where μ is a diffusion coefficient. The last term represents a macroscopic volatility—proportional to the parameter K—in which dB_t is a noise term that may not be white, i.e., the quadratic variation $[B_t, B_t]$ may not be proportional to t. While details will depend on the particular circumstances, such systems are subject to a distressingly rich spectrum of possible instabilities (Khasminskii 2012). The full set of equations would involve properly indexed sums across the parameters making up the vector $\mathbf{K}$.

A simple example. If a system following Eq. (8.19) has been initially placed in a characteristic eigenmode—e.g., the smooth part of a "fundamental diagram" flow on some traffic network—then the dynamic equation for deviation of a parameter $K(t)$ from that mode can be written, in first order, as

$$dK_t \approx aK_t dt + \sigma K_t dW_t \tag{8.20}$$

where dW_t represents white noise having uniform spectrum. Then, using the Ito chain rule on $\log[K]$,

$$d\log[K_t] \approx (a - \sigma^2/2)dt + \sigma dW_t \tag{8.21}$$

The expectation is then

$$E[K_t] \propto \exp[(a - \sigma^2/2)t] \tag{8.22}$$

so that, if $a < \sigma^2/2$, $E[K_t] \rightarrow 0$. σ^2 then—quite counterintuitively as described in Wallace (2016a)—is a kind of control information in the sense of the Data Rate Theorem that serves to stabilize system dynamics.

For an inherently unstable traffic flow, we impose closure on the model by taking $\sigma^2/2 = \mathcal{T}(\Gamma)$ so that higher "temperature" means more "noise," an intuitive result. Then, antiparalleling the arguments of Belletti et al. (2015, Sect. 2.3), for this simple example a "traffic Froude number" (TFN) $\mathcal{F}$ that defines regimes of free and turbulent flow, can be defined as

$$\mathcal{F} \equiv 1 - [a - \mathcal{T}] \tag{8.23}$$

where a is then a_0 in Eq. (8.5).

When $\mathcal{F} > 1$, the system is in "laminar" free-flow, and becomes "turbulent" when $\mathcal{F} < 1$.

A different characterization, from this perspective, is that $\mathcal{T}$ represents a kind of viscosity index so that $\mathcal{F}$ is more akin to a Reynolds number than to a classical Froude number.

A contrast between our approach and that of Belletti et al. (2015) lies in the central object-of-interest. They invoke a hydrodynamic perspective involving the "flow" of individual vehicles in a channel that finds "instability" to be associated with unconstrained travel speed. The focus here is on the stability of geodesics in the complex topological quotient space $\mathcal{M}^{2n}/W(r)$. This is, in a sense, the inverse of their problem.

As argued, raising Γ lowers $\mathcal{T}$, "freezing" the system from "liquid flow" to "crystallized" broken symmetries—platoons, shock fronts, jams, and myriad other "snowflake" structures that may include large-scale system-wide "lock-in" as described in Sects. 4.3 and 5.9 above.

8.8 Reconsidering Network Flow

The analysis of traffic flow failure on a road network is, conceptually, somewhat similar to characterizing the propagation of a signal via the Markov "network dynamics" formalism of Wallace (2016a) or Gould and Wallace (1994). This is an approach that might be used to empirically identify geodesic eigenmodes of real road network systems under different conditions, as opposed to individual vehicle dynamics or flow on a single road.

Following Gould and Wallace (1994), the spread of a "signal" on a particular network of interacting sites—between and within—is described at nonequilibrium steady state in terms of an equilibrium distribution ϵ_i "per unit area" A_i of a Markov process, where A scales with the different "size" of each node, taken as distinguishable by a scale variable A (for example, number of entering streets or average total traffic flow) as well as by its "position" i or the associated probability-of-contact matrix (POCM). The POCM is normalized to a stochastic matrix $\mathbf{Q}$ having unit row sums, and the vector ϵ calculated as $\epsilon = \epsilon\mathbf{Q}$

There is a vector set of dimensionless network flows $\mathcal{X}_t^i$, $i = 1, \ldots, n$ at time t. These are each determined by some relation

$$\mathcal{X}_t^i = g(t, \epsilon_i/A_i) \tag{8.24}$$

Here, i is the index of the node of interest, $\mathcal{X}_t^i$ is the corresponding dimensionless scaled ith signal, t the time, and g an appropriate function. Again, ϵ_i is defined by the relation $\epsilon = \epsilon\mathbf{Q}$ for a stochastic matrix $\mathbf{Q}$, calculated as the network probability-of-contact matrix between regions, normalized to unit row sums. Using $\mathbf{Q}$, we have broken out the underlying network topology, a fixed between-and-within travel configuration weighted by usage that is assumed to change relatively slowly on the timescale of observation compared to the time needed to approach the nonequilibrium steady state distribution.

Since the $\mathcal{X}$ are expressed in dimensionless form, g, t, and A must be rewritten as dimensionless as well giving, for the monotonic increasing (or threshold-triggered) function F

$$\mathcal{X}_\tau^i = G\left[\tau, \frac{\epsilon_i}{A_i} \times \mathcal{A}_\tau\right] \tag{8.25}$$

where $\mathcal{A}_\tau$ is the value of a "characteristic area" variate that represents the spread of the perturbation signal—evolving into a traffic jam under worst-case conditions—at (dimensionless) characteristic time $\tau = t/T_0$.

G may be quite complicated, including dimensionless "structural" variates for each individual geographic node i. The idea is that the characteristic "area" $\mathcal{A}_\tau$ grows according to a stochastic process, even though G may be a deterministic mixmaster driven by systematic local probability-of-contact or flow patterns. Then the appropriate model for $\mathcal{A}_\tau$ of a spreading traffic jam becomes something like Eq. (8.20), with K replaced by $\mathcal{A}$ and t by τ. Thus, for the network, the signal Y_τ must again have a "noise"/vehicle density threshold condition like Eq. (8.23) for large-scale propagation of a traffic jam across the full network—something that would look very similar to the spread of a power blackout.

Zhang (2015) uses a similar Markov method to examine taxicab GPS data for transit within and between 12 empirically identified "hot zones" in Shanghai, determining the POCM and its equilibrium distribution.

This approach is something in the spirit of a long line of work summarized by Cassidy et al. (2011) that attempts to extend the idea of a fundamental diagram for a single road to a full transport network. As they put it,

> Macroscopic fundamental diagrams (MFDs)...relate the total time spent to the total distance traveled...It is proposed that these macrolevel relations should be observed if the data come from periods when all lanes on all links throughout the network are in either the congested or the uncontested regime...

Following our arguments here, such conditions might apply when $\mathcal{A}_\tau \to 0$, or when it encompasses the entire network domain. Indeed, Fig. 8.3 suggests why MFDs cannot be constructed in general: Congested and free flowing sections of traffic networks will often, and perhaps usually, coexist in an essentially random manner depending on local traffic densities. Figure 8.6, adapted from Geroliminis and Sun (2011), shows the limitations of the MFD approach. It examines the flow, in vehicles/5 min intervals, vs. percent occupancy over a 3 day period for the Minnesota Twin Cities freeway network that connects St. Paul and Minneapolis. See Fig. 1 of their paper for details of the road and sensor spacing. Evidently, while the unconstrained region of occupancy permits characterization of a geodesic mode, both strong hysteresis and phase transition effects are evident after about 8% occupancy, analogous to the "nucleation" dynamics of Fig. 8.1 at high traffic density. Again, as in Fig. 8.1 "fine structure" should be expected within both geodesic and turbulent modes, depending on local parameters.

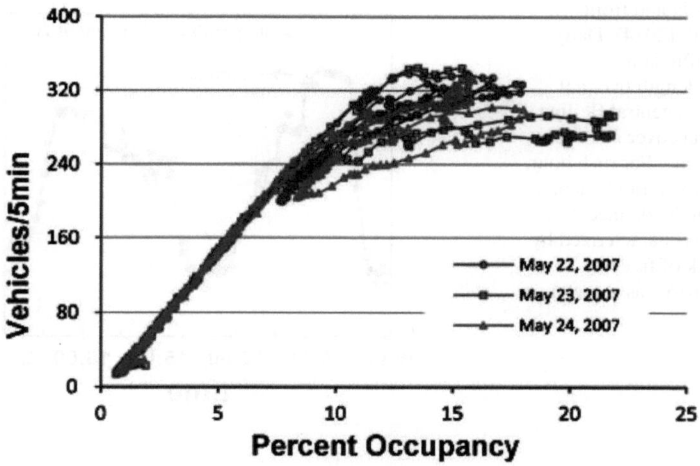

Fig. 8.6 Adapted from Geroliminis and Sun (2011). Breakdown of the macroscopic fundamental diagram for the freeway network connecting St. Paul and Minneapolis at high vehicle densities. Both nucleation and hysteresis effects are evident, showing the fine structure within the turbulent mode. As in Fig. 8.1c, breakdown begins near 8–10% occupancy

Daganzo et al. (2010) further find that MFD flow, when it can be characterized at all, will become unstable if the average network traffic density is sufficiently high. They find that, for certain network configurations, the stable congested state

> ...is one of complete gridlock with zero flow. It is therefore important to ensure that in real-world applications that a network's [traffic] density never be allowed to approach this critical value.

Daqing et al. (2014) examine the dynamic spread of traffic congestion on the Beijing central road network. They characterize the failure of a road segment to be a traffic velocity less than 20 km/h and use observational data to define a spatial correlation length in terms of the Euclidean distances between failed nodes. Our equivalent might be something like $\sqrt{\mathcal{A}_\tau}$. Adapting their results, Fig. 8.7 shows the daily pattern of the correlation length of cascading traffic jams over a 9 day period. The two commuting maxima are evident, and greatest correlation lengths reach the diameter of the main part of the city. Even at rush hour, no MFD can be defined, as, according to Fig. 8.3, the network will be a dynamic patchwork of free and congested components.

A next step would be to allow ρ, or a more general Γ, to vary in space and time, i.e., to parameterize the model using the moments of various density indices.

Figure 8.8, adapted from The Rand Fire Project (1979, Fig. 6.4), provides a disturbing counterexample to these careful empirical and theoretical results on network traffic flow, one with unfortunate results. Summarizing observations carried out by the Rand Fire Project, it represents a repeated sampling of "travel time vs.

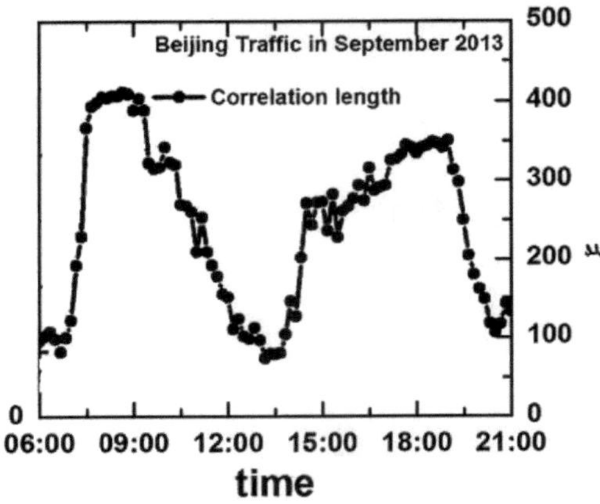

Fig. 8.7 Adapted from Daqing et al. (2014). Daily cycle of traffic jam correlation length over a 9 day period in central Beijing. The maxima cover most of the central city. For rush hour, no macroscopic fundamental diagram can be defined since the region is characterized by a patchwork of free and congested parts, as shown in Fig. 8.3

distance" for the full Trenton NJ road network in 1975 under varying conditions of time-of-day, day-of-week, weather, and so on, by fire companies responding to calls for service. This was an attempt to create a Macroscopic Fundamental Diagram in the sense used above, but without any reference at all to traffic density.

Indeed, fire service responses are a traffic flow "best case" as fire units are permitted to bypass one-way restrictions, traffic lights, and so on, and usually able to surmount even the worst weather conditions. In spite of best-case circumstances, the scatterplot evidently samples whole-network turbulent flow, not unlike that to the right of the local geodesic in Figs. 8.1 and 8.6, part of a single street and a highway network, respectively, and consistent with the assertions of Cassidy et al. (2011) that MFD relations can only be defined under very restrictive conditions, i.e., either complete free flow or full network congestion.

The Rand Fire Project, when confronted with intractable whole-network traffic turbulence, simply collapsed the data onto a "square root-linear" relation, as indicated on the figure. The computer models resulting from this gross oversimplification were used to determine fire service deployment strategies for high fire incidence, overcrowded neighborhoods in a number of US cities, with literally devastating results and consequent massive impacts on public health and public order. Wallace and Wallace (1998), produced under an Investigator Award in Health Policy Research from the Robert Wood Johnson Foundation, document the New York City case history. The Rand models are still in use by the New York City Fire Department, for political purposes outlined in that analysis.

8.9 Directed Homotopy

The symmetry breaking phase transition argument of Sect. 8.6 can be rephrased in terms of "directed homotopy"—dihomotopy—groupoids on an underlying road network, again parameterized by the "temperature" index $\mathcal{T}$. Classical homotopy characterizes topological structures in terms of the number of ways a loop within

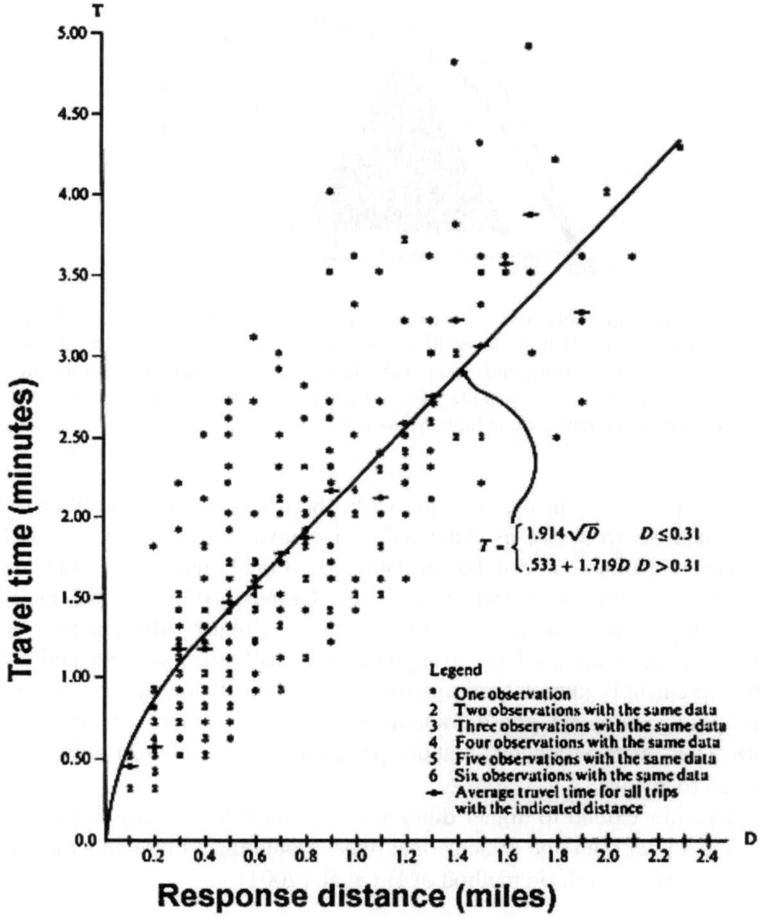

Fig. 8.8 Adapted from Fig. 6.4 of The Rand Fire Project (1979). Relation between fire company travel time and response distance for the full Trenton, NJ road network, 1975. The Rand Fire Project collapsed evident large-scale traffic turbulence into a simple "square root-linear" model used to design fire service deployment policies in high fire incidence, high population density neighborhoods of many US cities, including the infamous South Bronx. The impacts were literally devastating (Wallace and Wallace 1998)

the object can be continuously reduced to a base point (Hatcher 2001). For a sphere, all loops can be reduced. For a toroid—a donut shape—there is a hole so that two classes of loops cannot be reduced to a point. One then composes loops to create the "fundamental group" of the topological object. The construction is standard. Vehicles on a road network, however, are generally traveling from some initial point *So* to a final destination *S*1, as in Fig. 8.9, and directed paths, not loops are the "natural" objects, at least over a short time period, as in commuting.

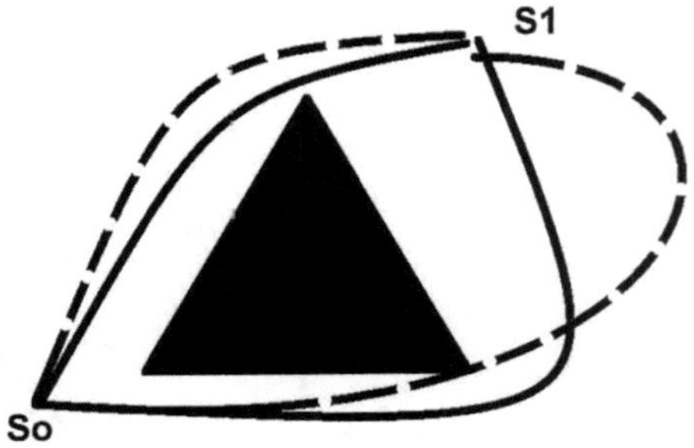

Fig. 8.9 Two equivalence classes of deformable paths connect the origin *So* with the destination *S*1, defining a groupoid. At high $\mathcal{T}$, both sets of paths are available for travel. At some point the synergism between crowding and road conditions creates a blockage in one or the other routes around the triangular "hole," breaking the groupoid symmetry. The "order parameter" is the number of blockages, which becomes zero at high symmetries

Given some "hole" in the road network, there will usually be more than one way to reach *S*1 from *So*, as indicated. An equivalence class of directed paths is defined by paths that can be smoothly deformed into one another without crossing barrier zones (Fajstrup et al. 2016; Grandis 2009), as indicated in the figure. At high values of the composite index $\mathcal{T}$, many different sets of paths will be possible, creating a large groupoid. As $\mathcal{T}$ declines, roadways and junctions become increasingly jammed, eliminating entire equivalence classes, and lowering the groupoid symmetry: phase transitions via classic symmetry breaking on a network. The "order parameter" that disappears at high $\mathcal{T}$ is then simply the number of blocked roadways.

These results extend to higher dihomotopy groupoids via introduction of cylindrical paths rather than one-dimensional lines, producing a more general version of the quotient space geodesic method of Hu et al. (2001).

8.10 Discussion and Conclusions

Ruelle (1983) raises a red flag that must apply to any traffic flow analysis:

> ...[A] deductive theory of developed turbulence does not exist, and a mathematical basis for the important theoretical literature on the subject is still lacking... A purely deductive analysis starting with the Navier–Stokes equation... does not appear feasible... and might be inappropriate because of the approximate nature of the... equation.

Or, as the mathematician Garrett Birkoff (1960, p. 5) puts it,

> ...[V]ery few of the deductions of rational hydrodynamics can be established rigorously.

Similar difficulties constrain the Black–Scholes models of financial engineering, and institutions that rely heavily on them have often gone bankrupt in the face of market turbulence (Wallace 2015c).

Turbulence in traffic flow does not represent simple drift from steady linear or even parallel travel trajectories. Traffic turbulence involves the exponential amplification of small perturbations into large-scale deviations from complicated streamline geodesics in a topologically complex map quotient space. This is the mechanism of groupoid "symmetry breaking" by which the system undergoes a phase transition from "liquid" geodesic flow to "crystalline" phases of shock fronts, platoons, and outright jams.

Under such circumstances, cognitive system initiative serves as a mechanism for returning to geodesic flows. Inhibition of cognitive initiative occurs when the composite density index Γ exceeds a critical limit, triggering complex dynamic condensation patterns and, for autonomous vehicle systems, perhaps even more disruptive behaviors.

It is, then, not enough to envision atomistic autonomous ground vehicles as having only local dynamics in an embedding traffic stream, as seems in the current American and European practice. Traffic light strategies, road quality, the usually rapid-shifting road map space, the dynamic composition of the traffic stream, bandwidth limits, and so on create the synergistic context in which single vehicles operate and which constitutes the individual "driving experience." It is necessary to understand the dynamics of that full system, not simply the behavior of a vehicle atom within it. The properties of that system will be both overtly and subtly emergent, as will, we assert, the responses of cognitive vehicles enmeshed in context, whether controlled by humans or computers.

One inference from this analysis is that failure modes afflicting large-scale V2V/V2I systems are likely to be more akin to power blackouts than to traffic jams as we know them, and the description by Kinney et al. (2005) is of interest:

> Today the North American power grid is one of the most complex and interconnected systems of our time, and about one half of all domestic generation is sold over ever-increasing distances on the wholesale market before it is delivered to customers... Unfortunately the same capabilities that allow power to be transferred over hundreds of miles also enable the propagation of local failures into grid-wide events... It is increasingly recognized that understanding the complex emergent behaviors of the power grid can only be understood from a systems perspective, taking advantage of the recent advances in complex network theory...

Dobson (2007) puts it as follows:

> [P]robabalistic models of cascading failure and power system simulations suggest that there is a critical loading at which expected blackout size sharply increases and there is a power law in the distribution of blackout size... There are two attributes of the critical loading: 1. A sharp change in gradient of some quantity such as expected blackout size as one passes through the critical loading. 2. A power law region in probability distribution of blackout size at the critical loading. We use the terminology "critical" because this behavior is analogous to a critical phase transition in statistical physics.

Daqing et al. (2014), in fact, explicitly link traffic jams and power failures:

> Cascading failures have become major threats to network robustness due to their potential catastrophic consequences, where local perturbations can induce global propagation of failures… [that] propagate through collective interactions among system components…. [W]e find by analyzing our collected data that jams in city traffic and faults in power grid are spatially long-range correlated with correlations decaying slowly with distance. Moreover, we find in the daily traffic, that the correlation length increases dramatically and reaches maximum, when morning or evening rush hour is approaching…

While clever V2V/V2I management strategies might keep traffic streams in supercooled high-flow mode beyond critical densities, such a state is notoriously unstable, subject to both random and deliberately caused "condensation" into large-scale frozen zones. More subtle patterns of autonomous vehicle "psychopathology" may be even less benign, as studied in detail elsewhere (Wallace 2016b).

It is difficult to escape the inference that, despite understandable marketing hype and other wishful thinking, large-scale V2V/V2I autonomous vehicle systems may simply not be practical, particularly in a context of coupled social and infrastructure deterioration.

It has been said that "The language of business is the language of dreams." Business dreams, however, do not necessarily serve as a sound foundation for the design and implementation of public policies affecting the well-being of large populations.

Acknowledgements The author thanks Dr. D.N. Wallace for useful discussions.

References

Atlan, H., and I. Cohen. 1998. Immune information, self-organization, and meaning. *International Immunology* 10: 711–717.

Belletti, F., M. Huo, X. Litrico, and A. Bayen. 2015. Prediction of traffic convective instability with spectral analysis of the Aw-Rascle-Zhang model. *Physics Letters A* 379: 2319–2330.

Birkoff, G. 1960. *Hydrodynamics: A Study in Logic, Fact, and Similitude*, 2nd ed. Princeton: Princeton University Press.

Blandin, S., et al. 2011. A general phase transition model for vehicular traffic. *SIAM Journal of Applied Mathematics* 71: 107–127.

Cassidy, M., K. Jang, and C. Daganzo. 2011. Macroscopic fundamental diagrams for freeway networks. *Transportation Research Reports* 2260: 8–15.

Chiabaut, N. 2015. Evaluation of a multimodal urban arterial: the passenger macroscopic fundamental diagram. *Transportation Research B* 81: 410–420.

Cover, T., and J. Thomas. 2006. *Elements of Information Theory*, 2nd ed. New York: Wiley.

Daganzo, C., V. Gayah, and E. Gonzales. 2010. Macroscopic relations of urban traffic variables: An analysis of instability. Working Paper UCB-ITS-VWP-2010-4, UC Berkeley Center for Future Urban Transit.

Daqing, L., J. Yinan, K. Rui, and S. Havlin. 2014. Spatial correlation analysis of cascading failures: Congestions and blackouts. *Scientific Reports* 4: 5381.

de Groot, S., and P. Mazur. 1984. *Nonequilibrium Thermodynamics*. New York: Dover.

Dobson, I. 2007. Where is the edge for cascading failure?: Challenges and opportunities for quantifying blackout risk. IEEE Power Engineering Society General Meeting, Tampa, FL (dobson@engr.wisc.edu).

Fajstrup, L. E. Goubault, A. Mourgues, S. Mimram, and M. Raussen. 2016. *Directed Algebraic Topology and Concurrency*. New York: Springer.

Feynman, R. 2000. *Lectures on Computation*. New York: Westview Press.

Geroliminis, N., and J. Sun. 2011. Properties of a well-defined macroscopic fundamental diagram for urban traffic. *Transportation Research B* 45: 605–617.

Geroliminis, N., N. Zeng, and K. Ampountolas. 2014. A three-dimensional macroscopic fundamental diagram for mixed bi-modal urban networks, *Transportation Research C* 42: 168–181.

Glazebrook, J.F., and R. Wallace. 2009. Rate distortion manifolds as model spaces for cognitive information. *Informatica* 33: 309–345.

Gould, P., and R. Wallace. 1994. Spatial structures and scientific paradox in the AIDS pandemic. *Geographysica Annaler* 76B: 105–116.

Grandis, M. 2009. *Directed Algebraic Topology: Models of Non-reversible Worlds*. New York: Cambridge University Press.

Hatcher, A. 2001. *Algebraic Topology*. New York: Cambridge University Press.

Helbing, D. 2001. Traffic and related self-driven many-particle systems. *Reviews of Modern Physics* 73: 1067–1141.

Hu, J., M. Prandini, K. Johnasson, and S. Sastry. 2001. Hybrid geodesics as optimal solutions to the collision-free motion planning problem. In *HSCC 2001*, ed. M. Di Benedetto and A. Sangiovanni-Vincentelli. LNCS, vol. 2034, 305–318. London: Springer.

Jin, C., et al. 2013. Spontaneous phase transition from free flow to synchronized flow in traffic on a single-lane highway. *Physical Review E* 87: 012815.

Kerner, B., and S. Klenov. 2009. Phase transitions in traffic flow on multilane roads. *Physical Review E* 80: 056101

Kerner, B., M. Koller, S. Klenov, H. Rehborn, and M. Leibel. 2015. Empirical features of spontaneous and induced traffic breakdowns in free flow at highway bottlenecks. ArXiv:1502.02862v2 [physics.soc-ph].

Khasminskii, R. 2012. *Stochastic Stability of Differential Equations*, 2nd ed. New York: Springer.

Kinney, R., et al. 2005. Modeling cascading failures in the North American power grid. *European Physics Journal B* 46: 101–107.

Landau, L., and E. Lifshitz. 1987. *Fluid Mechanics*, 2nd ed. New York: Pergamon Press.

Li, D., B. Fu, Y. Wang, G. Lu, Y. Berezin, and H.E. Stanley. 2015. Percolation transition in dynamical traffic network with evolving critical bottlenecks. *PNAS* 112: 669–672.

Maerivoet, S., and B. De Moor. 2006. Data quality travel time estimation and reliability, Katholieke University Leuven 06-030.

Nair, G., F. Fagnani, S. Zampieri, and R. Evans. 2007. Feedback control under data rate constraints: An overview. *Proceedings of the IEEE* 95: 108–137.

Pettini, M. 2007. *Geometry and Topology in Hamiltonian Systems*. New York: Springer.

Protter, P. 1990. *Stochastic Integration and Differential Equations*. New York: Springer.

Ramaswamy, S. 2010. The mechanics and statistics of active matter. *Annual Reviews of Condensed Matter Physics* 1: 323–345.

Ruelle, D. 1983. Turbulent dynamical systems. In *Proceedings of the International Congress of Mathematicians*, Aug. 16–24, Warsaw, 271–286.

Sugiyama, Y., M. Fukui, M. Kikuchi, K. Hasebe, A. Nakayama, K. Nishinari, S. Tadaki, and S. Yukawa. 2008. Traffic jams without bottlenecks – Experimental evidence for the physical mechanism of the formation of a jam. *New Journal of Physics* 10: 033001.

The Rand Fire Project. 1979. *Fire Deployment Analysis: A Public Policy Analysis Case Study*. New York: North Holland.

Wallace, R. 2012. Consciousness, crosstalk, and the mereological fallacy: An evolutionary perspective. *Physics of Life Reviews* 9: 426–453.

Wallace, R. 2014. Extending Swerdlow's hypothesis: Statistical models of mitochondrial deterioration and aging. *Journal of Mathematical Chemistry* 52: 2663–2679.

Wallace, R. 2015a. Closed-system 'economic' models for psychiatric disorders: Western atomism and its culture-bound syndromes. *Cognitive Processing* 16: 279–290.

Wallace, R. 2015b. *An Information Approach to Mitochondrial Dysfunction*. Singapore: World Scientific.

Wallace, R. 2015c. *An Ecosystem Approach to Economic Stabilization: Escaping the Neoliberal Wilderness*. Routledge Advances in Heterodox Economics. London: Routledge.

Wallace, R. 2016a. Subtle noise structures as control signals in high-order biocognition. *Physics Letters A*. doi: 10.1016/j.physleta.2015.11.037.

Wallace, R. 2016b. *Information Theory Models of Instabilities in Critical Systems*. Singapore: World Scientific.

Wallace, D., and R. Wallace. 1998. *A Plague on Your Houses*. New York: Verso.

Wallace, R., and D. Wallace. 2013. *A Mathematical Approach to Multilevel, Multiscale Health Interventions: Pharmaceutical Industry Decline and Policy Response*. London: Imperial College Press.

Weinstein, A. 1996. Groupoids: Unifying internal and external symmetry. *Notices of the American Mathematical Society* 43: 744–752.

Zhang, L. 2015. Realistic, efficient and secure geographic routing in vehicular networks, Ph.D. Thesis in the Department of Computer Science, University of Victoria (CA).

Chapter 9
Psychopathologies of Automata II: Autonomous Weapons and Centaur Systems

Summary Powerful asymptotic limit theorems of control and information theories illuminate the dynamics of autonomous weapons and man/machine "centaur" or "cockpit" systems under increasing fog-of-war burdens. Adapting tools from previous chapters, a relatively simple analysis shows there will not be graceful degradation in targeting precision, but sudden, highly punctuated, collapse to a pathological state in which "all possible targets are enemies." A central focus is the complex structure of the fog-of-war ecosystem itself, resulting in statistical tools not unlike regression models that should be useful in data analysis.

9.1 Introduction

The catastrophic drone wars in the Middle East and Africa (Columbia 2012; Stanford/NYU 2012; Wallace 2016a) will haunt the USA well into the next century, just as the miscalculations that created and followed World War I—including the European colonial "country building" producing Iraq and Syria—haunt us today. The USA and others are poised to move beyond remotely piloted drone systems to autonomous weapons and/or centaur warfighting—enhanced man/machine "cockpits" (Scharre 2016). It is asserted that centaur systems "keeping the man in the loop" will not only outperform automatons, but will also constrain, somewhat, the horrors of war.

Contrary perspectives abound. As Archbishop Silvano Tomasi (2014) puts it,

> ..[T]he development of complex autonomous weapon systems which remove the human actor from lethal decision-making is short-sighted and may irreversibly alter the nature of warfare in a less humane direction, leading to consequences we cannot possibly foresee, but that will in any case increase the dehumanization of warfare.

As Scharre describes, however, the First Gulf War Patriot missile fratricides (Hawley 2006; Wallace 2016a) raise significant questions regarding the operational reliability of such systems under fog-of-war constraints. The Patriot can be seen as an early example of forthcoming centaur man/machine composites.

R. Wallace, *Computational Psychiatry*, DOI 10.1007/978-3-319-53910-2_9

Trsek (2014) examines the 1988 US AEGIS system downing of a civilian airliner from a similar perspective, concluding that

> [Command responsibility] is already several steps removed from the operator in practice—it is naive to believe that we are relying on biological sensing to fulfill [rules-of-engagement] criteria, where the majority of information is electronically derived.

To address some of these matters, we expand the approach of Wallace (2016a), who examined canonical failure modes of real-time control systems using insights from cognitive theory. That work viewed such failures from the general perspective of the Data Rate Theorem that links control and information theories. Here, using the approach of previous chapters, we explore in considerably finer detail the structure and dynamics of the fog-of-war constraints that can collapse such systems into a ground state pathology in which "all possible targets are enemies." Such collapse moves beyond Scharre's "operational risk" into violations of the Laws of Land Warfare that require distinction between combatants and non-combatants.

We review and extend the formal linkage between control and information theories, leading to deeper understanding of fog-of-war constraints.

9.2 The Data Rate Theorem

Unlike an aircraft that can remain in stable flight as long as the center of pressure is sufficiently behind the center of gravity, high-order cognitive systems like human sports and combat teams, man–machine "cockpits," self-driving vehicles, autonomous weapons systems, and modern fighter aircraft—built to be maneuverable rather than stable—operate in real time on rapidly shifting topological "highways" of complex multimodal demand. Facing these turbulent topologies, the cognitive system must receive a constant flow of sufficiently detailed information describing them. More prosaically, driving on a twisting, pot-holed road after dark and at high speed requires very good headlights.

It will become clear that high-order cognition is inherently unstable in the sense of the Data Rate Theorem (Nair et al. 2007), viewing incoming information about the rapidly shifting topology of demand as the control signal.

The Data Rate Theorem (DRT) states that there is a minimum rate at which control information must be provided for an inherently unstable system to remain stable. The most direct approach is a linear expansion near a nonequilibrium steady state. An n-dimensional vector of system parameters at time t, x_t, determines the state at time $t+1$ according to the model of Fig. 9.1, so that

$$x_{t+1} = \mathbf{A}x_t + \mathbf{B}u_t + W_t \tag{9.1}$$

$\mathbf{A}, \mathbf{B}$ are fixed $n \times n$ matrices, u_t is the vector of control information, and W_t is an n-dimensional vector of white noise. The Data Rate Theorem (DRT) under such

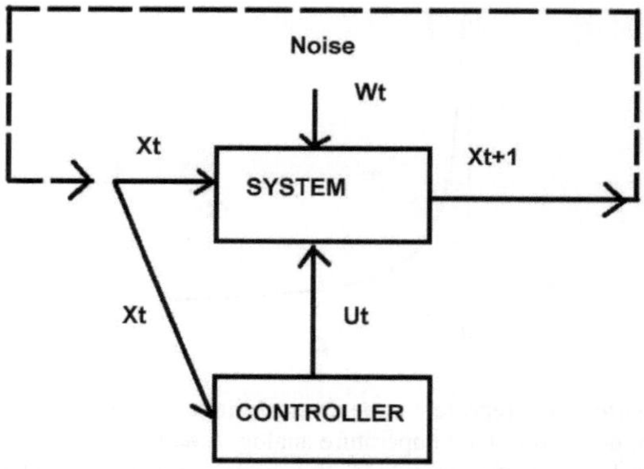

Fig. 9.1 A linear expansion near a nonequilibrium steady state of an inherently unstable control system, for which $x_{t+1} = \mathbf{A}x_t + \mathbf{B}u_t + W_t$. $\mathbf{A}, \mathbf{B}$ are square matrices, x_t the vector of system parameters at time t, u_t the vector of control signals at time t, and W_t a white noise vector. The Data Rate Theorem states that the minimum rate at which control information must be provided for system stability is $\mathcal{H} > \log[|\det[\mathbf{A}^m|]$, where $\mathbf{A}^m$ is the subcomponent of $\mathbf{A}$ having eigenvalues ≥ 1

conditions states that the minimum control information rate $\mathcal{H}$ is determined by the relation

$$\mathcal{H} > \log[|\det(\mathbf{A}^m)|] \equiv a_0 \tag{9.2}$$

where, for $m \leq n$, $\mathbf{A}^m$ is the subcomponent of $\mathbf{A}$ having eigenvalues ≥ 1. The right-hand side of Eq. (9.2) is interpreted as the rate at which the system generates "topological information." The Mathematical Appendix uses the Rate Distortion Theorem (RDT) to derive a more general version of this result.

The next step is both deceptively simple and highly significant. Given, for the moment, a scalar parameter ρ as an index of multimodal system demand, here representing the fog-of-war (most simply, for example, the magnitude of the dominant vector in a principal component analysis) we extend Eq. (9.2) as

$$\mathcal{H}(\rho) > f(\rho)a_0 \tag{9.3}$$

where a_0 is constant characteristic of low system demand and $f(\rho)$ is a positive, monotonically increasing function. The Mathematical Appendix uses a Black–Scholes model to approximate the "cost" of $\mathcal{H}$ as a function of the "investment" ρ. The first approximation is, surprisingly (or not), linear, so that $\mathcal{H} \approx \kappa_1\rho + \kappa_2$. Taking $f(\rho)$ to similar order, so that

$$f(\rho) \approx \kappa_3\rho + \kappa_4 \tag{9.4}$$

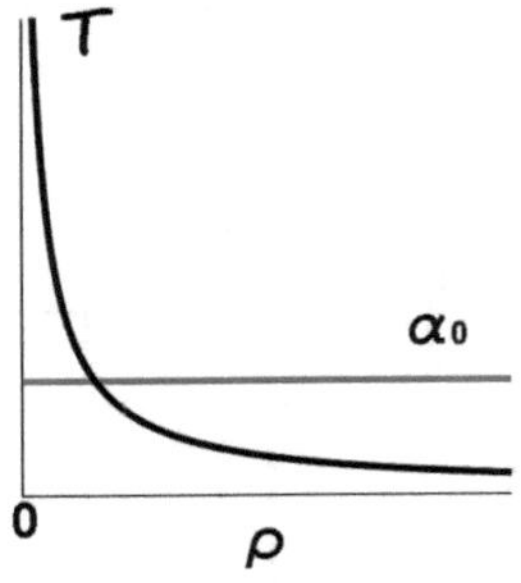

Fig. 9.2 The *horizontal line* represents the critical limit a_0. If $\kappa_2/\kappa_4 \gg \kappa_1/\kappa_3$, at some intermediate value of demand ρ, the temperature analog $\mathcal{T} \equiv (\kappa_1\rho + \kappa_2)/(\kappa_3\rho + \kappa_4)$ falls below that limit, the system becomes "supercooled," and sudden, punctuated, "phase transition" failure becomes increasingly probable

the limit condition for stability becomes

$$\mathcal{T}(\rho) \equiv \frac{\kappa_1\rho + \kappa_2}{\kappa_3\rho + \kappa_4} > a_0 \tag{9.5}$$

For $\rho = 0$, the stability condition is $\kappa_2/\kappa_4 > a_0$. At large ρ the condition becomes $\kappa_1/\kappa_3 > a_0$. If $\kappa_2/\kappa_4 \gg \kappa_1/\kappa_3$, the stability condition may be violated at high demand densities, and instability becomes manifest. See Fig. 9.2.

9.3 The Fog-of-War Ecosystem

The next stage of the argument is not simple. For a control system embedded in a complex and dynamic demand stream—here, the fog-of-war ecology—there will be several (many) ρ-values that are not independent but interact with each other. There is not, then, a simple scalar index of demand, but rather an $n \times n$ matrix $\hat{\rho}$ with elements $\rho_{i,j}$, $i, j = 1 \ldots n$. In general, as opposed to correlation matrices, $\rho_{i,j} \neq \rho_{j,i}$, since influences need not be symmetric.

Can there still be a single scalar "ρ" under such circumstances so that the conditions of Fig. 9.2 apply? An $n \times n$ matrix $\hat{\rho}$ has n invariants r_i, $i = 1 \ldots n$ that remain fixed when "principal component analysis" transformations are applied to data. The invariants are found using the famous characteristic equation

$$\mathcal{P}(\lambda) = \det(\hat{\rho} - \lambda I_n) = \lambda^n + r_1\lambda^{n-1} + \cdots + r_{n-1}\lambda + r_n \tag{9.6}$$

and the r_i can be used to construct an invariant scalar measure.

Here, det is the determinant, λ is a parameter that is an element of a ring, and I_n the $n \times n$ identity matrix. The invariants are the coefficients of λ in the polynomial $\mathcal{P}(\lambda)$, normalized so that the coefficient of λ^n is 1.

Since $n \times n$ matrices themselves form a ring, one has the classic relation

$$\mathcal{P}(\hat{\rho}) = 0 \times I_n \tag{9.7}$$

so that a matrix satisfies its own characteristic equation.

Again, an $n \times n$ matrix will have n such invariants from which a scalar index $\Gamma = g(r_1, \ldots, r_n)$ can be constructed to replace ρ in Eq. (9.5). Typically, the first invariant will be the trace and the last $\pm$ the determinant of $\hat{\rho}$.

The reduction of the complicated interaction matrix $\hat{\rho}$ to the scalar Γ is likely to be an ambiguous matter. We are making an approximation that must be fitted to each fog-of-war ecosystem: there will be no one-size-fits-all simplification methodology, although there may be equivalence classes of different systems that can be mapped onto a particular method.

This method can be seen as a variant of the Rate Distortion Manifold approach of Glazebrook and Wallace (2009) or the Generalized Retina used by Wallace and Wallace (2013, Sect. 10.1). Both find necessary conditions that high dimensional data flows can be projected down onto lower dimensional, shifting, tunable "tangent spaces" with minimal loss of essential information. Here, we ask that a complicated matrix interaction be projected onto a scalar function, a significant constraint that will limit the accuracy of the technique. Permitting a higher dimensional tangent space would improve possible fits to data, but would require significantly more formal overhead, and remains to be done.

Essentially we are defining $\mathcal{T}(\Gamma)$ as a synergistic "temperature" index characterizing fog-of-war conditions. Further analysis might uncover other "thermodynamic" quantities characterizing the combat ecosystem. For example, a more complete description of complex multimodal demand and its impacts on high-order, real-time cognitive systems—autonomous weapons or man/machine cockpit/centaurs—might involve concepts of "friction" in addition to fog-of-war "temperature." That is, we assume here that a weapon system can act promptly and without loss of effectiveness. Quite often, however, combat operations are confronted both by increasing actuation delays and progressive loss of efficiency under stress. The incorporation of delay in the kind of stochastic differential equation models we use here, however, introduces formidable mathematical complications, as would inclusion of stochastic models of damage accumulation.

The simplest approach would be to combine the fog-of-war "temperature" index with delay and attrition measures into a single "wind-chill" factor.

9.4 The Dynamics of Control Failure

What are the dynamics of $\mathcal{T}$ under stochastic circumstances? That is, how is a control signal u_t in Fig. 9.1 expressed in the system response x_{t+1}? We deterministically retranslate an observed sequence of system outputs $X^i = x_1^i, x_2^i, \ldots$ into a sequence of possible control signals $\hat{U}^i = \hat{u}^i_0, \hat{u}^i_1, \ldots$ and compare that sequence with the original control sequence $U^i = u^i_0, u^i_1, \ldots$, with the difference between them having a particular value under some chosen distortion measure and hence having an average distortion

$$< d >= \sum_i p(U^i) d(U^i, \hat{U}^i) \tag{9.8}$$

where $p(U^i)$ is the probability of the sequence U^i and $d(U^i, \hat{U}^i)$ is the distortion between U^i and the sequence of control signals that has been deterministically reconstructed from the system output.

It is then possible to apply a classic Rate Distortion argument (Cover and Thomas 2006). According to the Rate Distortion Theorem, there exists a Rate Distortion Function, $R(D)$, that determines the minimum channel capacity necessary to keep the average distortion below some fixed limit D (Cover and Thomas 2006). Based on Feynman's (2000) interpretation of information as a form of free energy, it is possible to construct a Boltzmann-like pseudoprobability density in the "temperature" $\mathcal{T}$ as

$$dP(R, \mathcal{T}) = \frac{\exp[-R/\mathcal{T}]dR}{\int_0^\infty \exp[-R/\mathcal{T}]dR} \tag{9.9}$$

since higher $\mathcal{T}$ must necessarily be associated with greater channel capacity.

The integral in the denominator can be interpreted as a statistical mechanical partition function, and it is possible to define a "free energy" Morse Function F (Pettini 2007) as

$$\exp[-F/\mathcal{T}] = \int_0^\infty \exp[-R/\mathcal{T}]dR = \mathcal{T} \tag{9.10}$$

so that $F(\mathcal{T}) = -\mathcal{T} \log[\mathcal{T}]$.

Then an entropy-analog can also be defined as the Legendre transform of F

$$\mathcal{S} \equiv F(\mathcal{T}) - \mathcal{T} dF/d\mathcal{T} = \mathcal{T} \tag{9.11}$$

The Onsager approximation to nonequilibrium thermodynamics (de Groot and Mazur 1984) can now be applied, using the gradient of $\mathcal{S}$ in $\mathcal{T}$, so that the dynamics of $\mathcal{T}$ are represented by a stochastic differential equation

$$\begin{aligned} d\mathcal{T}_t &= (\mu d\mathcal{S}/d\mathcal{T})dt + \beta \mathcal{T}_t dW_t \\ &= \mu dt + \beta \mathcal{T}_t dW_t \end{aligned} \tag{9.12}$$

where μ is a diffusion coefficient and β is the magnitude of the impinging white noise dW_t.

As above, in the context of Jensen's inequality for a concave function, applying the Ito chain rule to $\log(\mathcal{T})$ in Eq. (9.12) gives the familiar result for the nonequilibrium steady state expectation of $\mathcal{T}$ as

$$E(\mathcal{T}_t) \geq \frac{\mu}{\beta^2/2} \tag{9.13}$$

μ is interpreted as the attempt by the control apparatus—autonomous, centaur, cockpit—to maintain stability. Thus rising system noise can significantly increase the probability that $\mathcal{T}$ falls below the critical limit, triggering a punctuated failure of control.

It is important to recognize, however, that, since $E(\mathcal{T})$ is an expectation, in this model *there will always be some probability that* $\mathcal{T}(\Gamma)$ *will fall below the critical value* a_0.

Raising μ and limiting β decreases that probability, but cannot eliminate it: sudden onset of instability is always possible, triggering a "false target attack" in the sense of Kish et al. (2009):

> Unfortunately... adjusting the sensor threshold to increase the number of target attacks also increases the number of false target attacks. Thus the operator's objectives are competing, and a trade-off situation arises.

It is not difficult to construct numerical simulations of these results, for example, using the ItoProcess construct available in later versions of the computer algebra program Mathematica.

9.5 The Dynamics of High-Level Cognitive Dysfunction

The DRT argument above implies a raised probability of a transition between stable and unstable behavior if the temperature analog $\mathcal{T}(\Gamma)$ from Eq. (9.5) falls below a critical value, as in Fig. 9.2. We can extend the perspective to more complicated patterns of phase transition via the "cognitive paradigm" of Atlan and Cohen (1998), who view a system as cognitive if it compares incoming signals with a learned or inherited picture of the world, then actively chooses a response from a larger set of those possible to it. Choice implies the existence of an information source, since it reduces uncertainty in a formal way (Wallace 2012, 2015, 2016a).

Given such a "dual" information source associated with the inherently unstable cognitive system of interest, an equivalence class algebra can be constructed by choosing different system origin states and defining the equivalence of subsequent states at a later time by the existence of a high probability path connecting them to the same origin state. Disjoint partition by equivalence class, analogous to orbit equivalence classes in dynamical systems, defines a symmetry groupoid associated with the cognitive process (Wallace 2012). Groupoids are generalizations of group symmetries in which there is not necessarily a product defined for each possible element pair (Weinstein 1996). An example would be the disjoint union of different groups.

The equivalence classes across possible origin states define a set of information sources dual to different cognitive states available to the inherently unstable cognitive system. These create a large groupoid, with each orbit corresponding to a transitive groupoid whose disjoint union is the full groupoid. Each subgroupoid is associated with its own dual information source, and larger groupoids must have richer dual information sources than smaller.

Let X_{G_i} be the system's dual information source associated with groupoid element G_i. Given Eqs. (9.5)–(9.7), we construct a Morse Function (Pettini 2007) in a standard manner, using $\Gamma = g(r_1, \ldots, r_n)$ in Eq. (9.5) in place of ρ.

Let $H(X_{G_i}) \equiv H_{G_i}$ be the Shannon uncertainty of the information source associated with the groupoid element G_i. Define another Boltzmann-like pseudo-probability as

$$P[H_{G_i}] \equiv \frac{\exp[-H_{G_i}/\mathcal{T}]}{\sum_j \exp[-H_{G_j}/\mathcal{T}]} \tag{9.14}$$

where the sum is over the different possible cognitive modes of the full system.

Another "free energy" Morse Function $\mathcal{F}$ can then be defined as

$$\exp[-\mathcal{F}/\mathcal{T}] \equiv \sum_j \exp[-H_{G_j}/\mathcal{T}]$$

$$\mathcal{F} = -\mathcal{T} \log\left[\sum_j \exp[-H_{G_j}/\mathcal{T}]\right] \tag{9.15}$$

As a consequence of the underlying groupoid generalized symmetries associated with high-order cognition, as opposed to simple control theory, it is possible to apply an extension of Landau's version of phase transition (Pettini 2007). Landau argued that spontaneous symmetry breaking of a group structure represents phase change in physical systems, with the higher energies available at higher temperatures being more symmetric. The shift between symmetries is highly punctuated in the temperature index, here the "temperature" analog of Eq. (9.5), in terms of the scalar construct $\Gamma = g(r_1, \ldots, r_n)$, but in the context of groupoid rather than group symmetries.

Based on the analogy with physical systems, there should be only a few possible phases, with highly punctuated transitions between them as the fog-of-war "temperature" $\mathcal{T}$ decreases, ultimately freezing the operational phenotype into the usual always-on mode: "Kill everything, and let God sort them out." Sufficient conditions for the stability of this pathological state are discussed in Sects. 4.3 and 5.9 above.

9.6 Discussion and Conclusions

According to the Data Rate Theorem, if the rate at which control information can be provided to an unstable system is below the critical limit defined by the rate at which the system generates "topological information," there is no coding strategy, no timing strategy, no control scheme of any form, that can ensure stability. Generalization to the rate of incoming information from the rapidly changing multimodal "roadway" environments in which a real-time cognitive system must operate suggests that there will be sharp onset of serious dysfunction under the burden of rising demand. Here we have analyzed that multimodal demand in terms of the crosstalk-like fog-of-war matrix $\rho_{i,j}$ that can be characterized by situation-specific statistical models leading to the scalar temperature analog $\mathcal{T}$. More complicated "tangent space" reductions are possible, at the expense of greater mathematical overhead (e.g., Glazebrook and Wallace 2009).

There will not be graceful degradation under falling fog-of-war "temperature" or "wind-chill" factors, but rather punctuated functional decline that, for autonomous, centaur, or man–machine cockpit weapon systems, deteriorates into a frozen state in which "all possible targets are enemies," as in the case of the Patriot missile fratricides (Hawley 2006; Wallace 2016a). Other cognitive systems display analogous patterns of punctuated collapse into simplistic dysfunctional phenotypes or behaviors (Wallace 2015, 2016b): the underlying dynamic is ubiquitous and, apparently, inescapable. In sum, there is no free lunch for cognitive weapon systems, with or without hands-on human control. All such systems are inherently susceptible to serious operational instabilities under complex fog-of-war environments. Policy based on the business dreams of military contractors and their academic or think-tank clients—promises of precision targeting—will be confronted by nightmare realities of martyred civilian populations, recurring generations of new "terrorists," and the persistent stench of war crime.

References

Atlan, H., and I. Cohen. 1998. Immune information, self-organization, and meaning. *International Immunology* 10: 711–717.

Columbia University Law School Human Rights Clinic. 2012. Counting Drone Strike Deaths. http://web.law.columbia.edu/human-rights-institute.

Cover, T., and J. Thomas. 2006. *Elements of Information Theory*, 2nd ed. New York: Wiley.

de Groot, S., and P. Mazur. 1984. *Nonequilibrium Thermodynamics*. New York: Dover.

Feynman, R. 2000. *Lectures on Computation*. New York: Westview Press.

Glazebrook, J.F., and R. Wallace. 2009. Rate distortion manifolds as model spaces for cognitive information. *Informatica* 33: 309–345.

Hawley, J. 2006. Patriot Fratricides: The human dimension lessons of Operation Iraqi Freedom. *Field Artillery*, January–February.

Kish, B., M. Pachter, and D. Jacques. 2009. Effectiveness measures for operations in uncertain environments. In *UAV Cooperative Decision and Control: Challenges and Practical Applications*, ed. T. Shima and S. Rasmussen, Chap. 7. Philadelphia: SIAM Publications.

Nair, G., F. Fagnani, S. Zampieri, and R. Evans. 2007. Feedback control under data rate constraints: An overview. *Proceedings of the IEEE* 95: 108–137.

Pettini, M. 2007. *Geometry and Topology in Hamiltonian Dynamics*. New York: Springer.

Scharre, P. 2016. Autonomous weapons and operational risk. Center for New American Security, Washington, DC. http://www.cnas.org/autonomous-weapons-and-operational-risk.vtISHGO-RiY.

Stanford/NYU. 2012. Living under drones: Death, injury and trauma to civilians from US drone practices in Pakistan. http://livingunderdrones.org/.

Tomasi, S. 2014. Catholic Herald. http://www.catholicherald.co.uk/news/2014/05/15/vatican-official-voices-opposition-to-automated-weapons-systems/.

Trsek, R. (Lt. Col. USAF). 2014. Cutting the cord: Discrimination and command responsibility in autonomous lethal weapons. Available online from the Air War College of Air University, USA.

Wallace, R. 2012. Consciousness, crosstalk and the mereological fallacy: An evolutionary perspective. *Physics of Life Reviews* 9: 426–453.

Wallace, R. 2015. *An Information Approach to Mitochondrial Dysfunction: Extending Swerdlow's Hypothesis*. Singapore: World Scientific.

Wallace, R. 2016a. Canonical failure modes of real-time control systems: Insights from cognitive theory. *International Journal of Systems Science* 47: 1280–1295.

Wallace, R. 2016b. Canonical instabilities of autonomous vehicle systems. https://peerj.com/preprints/1714/.

Wallace, R., and D. Wallace. 2013. *A Mathematical Approach to Multilevel, Multiscale Health Interventions: Pharmaceutical Industry Decline and Policy Response*. London: Imperial College Press.

Weinstein, A. 1996. Groupoids: Unifying internal and external symmetry. *Notices of the American Mathematical Society* 43: 744–752.

Chapter 10
The Dynamics of Environmental Insult

Summary The principal environment of humans, their machines, man/machine composites, and institutions, is other entities of the same sort, creating an inherent positive feedback for the environmental insults triggering developmental and cognitive dysfunctions at different scales and levels of organization. We bookend the evolutionary discussion of the opening chapter on consciousness with an analogous treatment adapted from evolutionary economics, parallel to the resilience theory of ecosystem studies.

10.1 Introduction

Environmental insult plays a central role in developmental and cognitive dysfunction, affecting regulatory decisions at the directed homotopy branch points that characterize both ontology and decision. But, for humans, their machines and man/machine composites, and for institutions, the principal determinant, and often the major component, of that environment is those very things. This is, then, inherently a highly reflexive system that has long been characterized by evolutionary economists as "Schumpeterian" (e.g., Hodgson and Knudsen 2010; Wallace 2015 and the references therein). Such systems are subject to evolutionary selection pressures similar to, but different from, those outlined in the first chapter of this book. Nonetheless, the similarities are sufficient to permit formal analysis of the dynamics of environmental insult from an evolutionary perspective, leading to a "punctuated equilibrium" treatment analogous to the "domain shifts" of ecosystem resilience theory (Gould 2002; Holling 1973, 1992).

For the USA, two recent phenomena have driven large-scale patterns of environmental insult, triggering behavioral dysfunctions across multiple scales and levels of organization. These phenomena are widespread deurbanization and "rust belt" deindustrialization, both related to public policies and economic decisions (Wallace and Wallace 1998; Wallace 2015; Ullmann 1988; Melman 1971).

Beginning in the 1970s, cities like New York, Newark, Detroit, Cleveland, and elsewhere suffered outbreaks of contagious urban decay and building abandonment leaving vast areas resembling the bombed-out cities of Germany after WW II. In New York City, the outbreak was driven by cuts in fire-related municipal services delivered to minority voting blocs (Wallace and Wallace 1998). Figure 10.1 shows the loss of occupied housing units in the Bronx section of New York City

R. Wallace, *Computational Psychiatry*, DOI 10.1007/978-3-319-53910-2_10

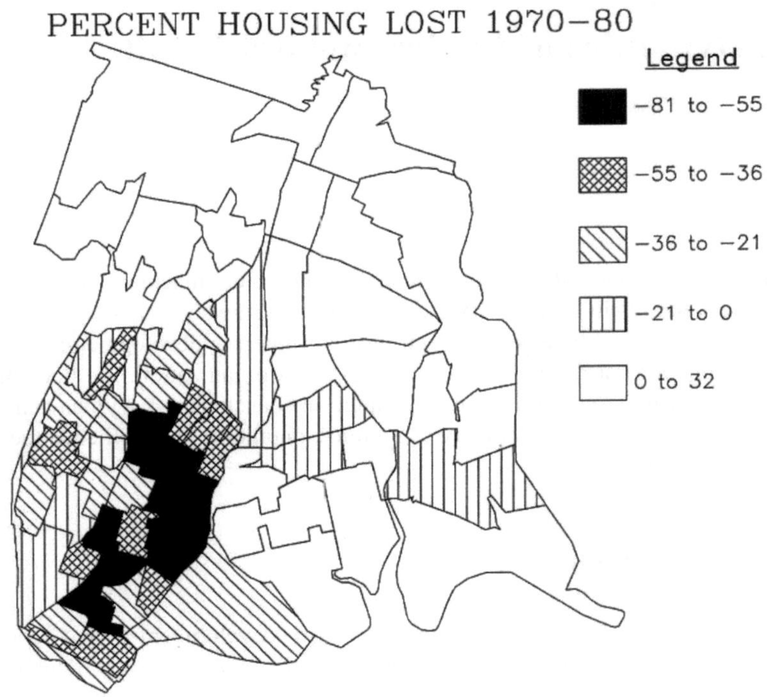

Fig. 10.1 Percent change in occupied housing units for New York City's Borough of the Bronx between 1970 and 1980. The degree of housing loss is unprecedented for an industrialized country outside of wartime. Similar losses have affected many US central cities, particularly across the "rust belt" counties suffering Cold War-induced deindustrialization (Wallace 2015, Chap. 7)

between 1970 and 1980. The Bronx, with 1.4 million people, is one of the largest conurbations in the Western world. Some districts lost as much as 80% of their housing units in this time, causing massive population displacement and loss of the social cohesion that is the basis of population physical and mental health (Wallace and Wallace 1998).

Widespread deindustrialization followed upon the death of the US economic "Red Queen" that had been fueled by a necessarily relentless focus of scientific and engineering resources on the improvement of civilian industry (Melman 1971; Ullmann 1988; Wallace 2015). That is, to remain globally competitive, Western "market" economies must improve productivity some 3% annually. Forever. Diversion of inherently limited scientific and technological resources into weapons development and production, what Seymour Melman called the establishment of "Pentagon capitalism," fatally undercut the ability of the USA to compete with "Asian tiger" countries.

Figure 10.2 shows the result: counties of the US rust belt losing 1000 or more industrial jobs between 1972 and 1987, well before the "globalization" policies now

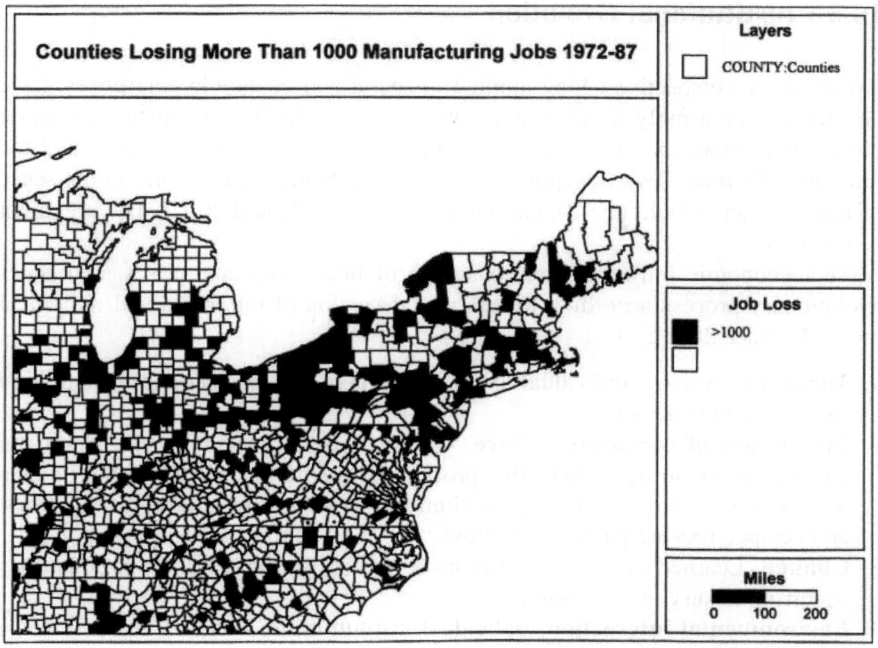

Fig. 10.2 "Rust Belt" counties of the USA losing 1000 or more industrial jobs between 1972 and 1987, well before formal "globalization" policies

blamed for the loss of those jobs. Globalization simply involved easing the use of offshore factories to manufacture what the USA could no longer competitively build at home.

Chapter 7 of Wallace and Wallace (2017) provides a more complete treatment of the public health effects of deindustrialization, and Chap. 7 of Wallace (2015) models the economic impact of the Cold War diversion of scientific and engineering resources from civilian to military enterprise in terms of a dynamic "Pentagon ratchet," treated in more detail here using evolutionary perspectives.

The evolutionary theory of institutional economics provides tools for understanding the action of selection pressure on institutional dynamics, and, at the population level for a Western nation, these dynamics determine levels of environmental insult resulting in the developmental disorders studied in the previous chapters. Even cognitive machine systems may be driven to dysfunction by the positive feedback of self-referential environmental insult. For example, under combat conditions it is easy to imagine a feedback loop in which lowering the "wind-chill" fog-of-war index $\mathcal{T}$ serves to intensify selection pressures that ratchet it further down.

We begin with a recapitulation of basic ideas from evolutionary economics.

10.2 Institutional Evolution

Evolutionary perspectives, long applied to social and economic enterprises, have become an increasingly attractive alternative to current failing "atomistic" economic theory (e.g., Hodgson and Knudsen 2010; Wallace 2013, 2015, Chap. 1, and the references therein). Two dynamics dominate: punctuated equilibrium, in the sense of Eldredge and Gould (1972), and path dependence (Gould 2002). The argument is as follows.

Socioeconomic infrastructure, consisting of interacting enterprises, undergoes evolutionary process according to a modified version of the traditional biological mode (Wallace 2013):

1. **Variation**. Among individual institutions there is considerable variation in structure and behavior.
2. **Inheritance of corporate culture**. Within its developmental path, which can be seen as a kind of reproductive process, an institution will resemble its own history more than that of other institutions, as corporate strategies, resources, and perspectives are passed on in time.
3. **Change**. Learned or enforced variation in structure and policy is constantly occurring in surviving institutions.
4. **Environmental interaction**. Individual institutions and related groups engage in powerful, often punctuated, dynamic mutual relations with their embedding environments that may include the exchange of "heritage material" between markedly different entities through learning, or the abduction or diffusion of ideas and opinions.

Many of the essential processes within this structure can be represented in terms of interacting information sources, constrained by the asymptotic limit theorems of information and control theories. Following the arguments of Wallace (2013, 2014, 2015), it can be shown that

1. An embedding ecosystem has "grammar" and "syntax" that allows it to be represented as an information source, say X.
2. Like genetic heritage, institutional heritage is also characterized as a "language," and hence an information source Y.
3. As described above, institutional cognition involves a dual information source, Z. Further, cognition is always associated with groupoids that generalize the idea of a symmetry group.
4. Large deviations in dynamical systems occur with very high probability only along certain developmental pathways, allowing definition of an information source we will call L_D. See Wallace (2013, 2014, 2015) for details that follow the arguments of Champagnat et al. (2006).

As a consequence, we can define a joint Shannon uncertainty representing the interaction of these information sources as

$$H(X, Y, Z, L_D)$$

Defining an "entropy" across a vector of system parameters **J** as the Legendre transform

$$S \equiv H(\mathbf{J}) - \mathbf{J} \cdot \nabla_{\mathbf{J}} H \tag{10.1}$$

we can apply, in first order, an analog to the Onsager approximations of earlier chapters. That is, a first order dynamic equation follows using the stochastic version of the Onsager formalism from nonequilibrium thermodynamics (de Groot and Mazur 1984)

$$dJ_t^i \approx \left(\sum_k \mu_{i,k} \partial S / \partial J_t^k \right) dt + \sigma_i J_t^i dB_t \tag{10.2}$$

$\mu_{i,k}$ defines a diffusion matrix, the σ_i are parameters, and dB_t represents a noise that may not the usual Brownian motion under undifferentiated white noise.

Setting the expectation of this relation to zero, we find a relatively large set of nonequilibrium steady states (nss), indexed by j. Each nss is characterized by an uncertainty value H_j.

Importing the environmental stress index $\mathcal{T}$ from earlier chapters,

$$\mathcal{T} = \frac{\kappa_1 \Gamma + \kappa_2}{\kappa_3 \Gamma + \kappa_4}$$

where Γ represents a scalar constructed from the invariants of a complex environmental crosstalk matrix, we can write a pseudoprobability for state q as

$$P_q = \frac{\exp(-H_q/\mathcal{T})}{\sum_j \exp(-H_j/\mathcal{T})} \tag{10.3}$$

and define a "free energy" Morse Function $\hat{F}$ in terms of the denominator sum, i.e.,

$$\exp(-\hat{F}/\mathcal{T}) \equiv \sum_j \exp(-H_j/\mathcal{T}) \tag{10.4}$$

Changes in $\mathcal{T}$ (inverse in the index of environmental insult Γ) will be associated with profound—and highly punctuated—evolutionary transitions (Eldredge and Gould 1972; Gould 2002; Wallace 2014). These transitions, involving cognitive groupoid analogs to physical "symmetry breaking" (Pettini 2007), then define entirely new pathways along which socioeconomic systems develop. There is never, ever, a "return to normal after perturbation" in path-dependent evolutionary process.

Taking the perspective of Wallace (2011), it is possible to examine an economic-like ratchet that, in evolutionary terms, is usually characterized as a "self-referential" dynamic. Goldenfeld and Wose (2010) describe the mechanism for biological evolution:

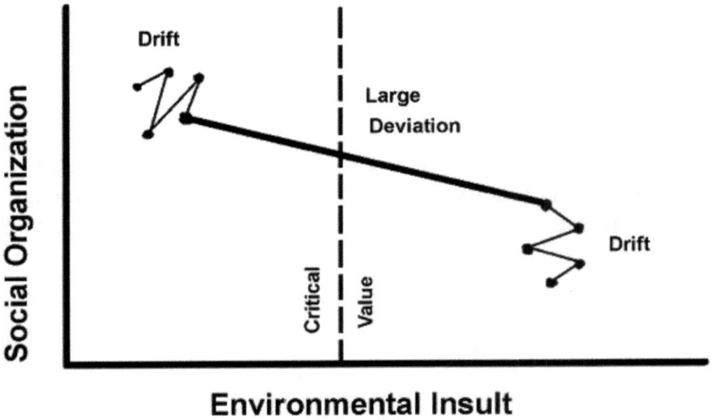

Fig. 10.3 The vertical axis indexes degree of social organization. The horizontal axis represents the degree of environmental insult Γ. At low insult the system drifts about a nonequilibrium steady state with significant social integration and organization. Insult exceeding some critical level triggers a punctuated phase change via a large deviation, leading to a less organized nonequilibrium steady state. Social disintegration, of itself, constitutes a serious environmental insult, leading to "self-referential" ratchet dynamics: a positive feedback-driven race to the bottom. Conversely, however, significant investment in social, technological, and physical infrastructure can trigger an upward ratchet

> ... [T]he genome encodes the information which governs the response of an organism to its physical and biological environment. At the same time, this environment actually shapes genomes through gene transfer processes and phenotype selection. Thus, we encounter a situation where the dynamics must be self-referential: the update rules change during the time evolution of the system, and the way in which they change is a function of the state and thus the history of the system... self-referential dynamics is an inherent and probably defining feature of evolutionary dynamics...

The evolutionary dynamic we propose for socioeconomic systems under the stress of environmental insult is illustrated by Fig. 10.3. The vertical axis represents an index of social organization—percent voting, educational attainment, percent active in civic associations or religious institutions, etc. The horizontal axis is taken as a measure of environmental stress Γ. At low levels of insult the system drifts about some nonequilibrium steady state having relatively high degrees of social integration. When stress exceeds a threshold, there is a punctuated phase change associated with a large deviation, leading to a less organized nonequilibrium steady state, as indicated. Thus onset of social disintegration itself constitutes a significant environmental insult, leading to a fully self-referential downward ratchet.

A relatively simple deterministic mathematical description of such a binary switch might be as follows. Assume Γ, the scalar index of environmental stress, is initially at some nonequilibrium steady state, and that $\Gamma \to \Gamma + \Delta$. Then Δ is assumed, in first order, to follow a relation

$$d\Delta/dt = \mu\Delta - C/\Delta, \ C, \mu > 0 \tag{10.5}$$

so that, if $\Delta \leq \sqrt{C/\mu}$, then $d\Delta/dt \leq 0$, and the system remains at or near Γ. Otherwise $d\Delta/dt$ becomes positive, and the switch is triggered, according to Fig. 10.3. Other models that lead to such quasi-stability could be used.

Next, we expand in a stochastic treatment, so that

$$d\Delta_t = (\mu\Delta_t - C/\Delta_t)dt + \sigma\Delta_t dW_t \tag{10.6}$$

where σ is an index of the magnitude of impinging white noise dW_t. Then, applying the Ito chain rule to $\log[\Delta_t]$ (Protter 1990), in the context of Jensen's inequality for a concave function, the nonequilibrium steady state expectation is

$$E[\Delta_t] \geq \sqrt{\frac{C}{\mu - \sigma^2/2}} \tag{10.7}$$

Sufficient noise drives an explosive perturbation. In addition, since Eq. (10.7) is an expectation across a probability distribution, even at relatively low mean values there may well be much larger stochastic excursions—large deviations—that can trigger a destabilizing transition, following Fig. 10.3. Again, Wallace (2015, Chap. 7) examines the impact of the diversion of technological resources from civilian to military industrial enterprise during the Cold War leading to the massive "rust belt" collapse in the USA.

Of course, given sufficient "available free energy," in a large sense, upward ratchets in levels of organization—analogous to the famous aerobic transition or, in human social systems, to the American Revolution, the Industrial Revolution, or the US Labor and Civil Rights Movements—are also possible.

10.3 Discussion and Conclusions

Given that the principal environment of humans, their machines, man/machine composites, and embedding institutions, is more of the same, it becomes clear that environmental insult driving both developmental and cognitive dysfunction can become a self-dynamic force, triggering outcomes that can fuel the fire, as it were. This can, as indicated, go both ways: ratchets can be upward as well as downward. Sufficient sociotechnical investment can initiate socioeconomic and cultural dynamics that reverse environmental factors that are driving developmental and cognitive dysfunctions across many scales and levels of organization. The precipitate decline in "normal" death rates for industrialized nations that following the reforms of the late nineteenth and early twentieth centuries provides a cogent case history.

Holling (1992) argues persuasively that, for ecosystems, mesoscale "keystone" processes in particular entrain both higher and lower levels of organization. Given the recent history of the USA, it seems clear that the essential mesoscale perturbations involved deindustrialization and deurbanization, individually and in synergism. Both dynamics were driven by public policy. Unlike ecosystems undergoing eutrophication, however, socioeconomic systems can be ratcheted upward. For the USA, both deindustrialization and deurbanization can, in theory, be reversed by public policy, as outlined by Ullmann (1988) and Wallace (2015, Sect. 7.6). In brief, reversing the American Catastrophe would require ending "Pentagon Capitalism," active reindustrialization and infrastructure reconstruction, rebuilding of cities to recapture economies of scale, public education through the Masters Degree for those who wish it, and draconian opposition to the scapegoating of minorities that is not only a distraction from real issues, but also the canonical process by which a dying empire eats its own entrails.

References

Champagnat, N., R. Ferriere, and S. Meleard. 2006. Unifying evolutionary dynamics: From individual stochastic processes to macroscopic models. *Theoretical Population Biology* 69: 297–321.

de Groot, S., and P. Mazur. 1984. *Nonequilibrium Thermodynamics*. New York: Dover.

Eldredge, N., and S. Gould. 1972. Punctuated equilibrium: An alternative to phyletic gradualism. In *Models in Paleobiology*, ed. T. Schopf, 82–115. San Francisco: Freeman, Cooper and Co.

Goldenfeld, N., and C. Wose. 2010. Life is physics: Evolution as a collective phenomenon far from equilibrium. ArXiv:1011.4125v1 [q-bio.PE].

Gould, S. 2002. *The Structure of Evolutionary Theory*. Cambridge MA: Harvard University Press.

Hodgson, G., and T. Knudsen. 2010. *Darwin's Conjecture: The Search for General Principles of Social and Economic Evolution*. Chicago, IL: University of Chicago Press.

Holling, C. 1973. Resilience and stability of ecological systems. *Annual Reviews of Ecological Systematics* 4: 1–23.

Holling, C. 1992. Cross-scale morphology, geometry and dynamics of ecosystems. *Ecological Monographs* 62: 447–502.

Melman, S. 1971. *The War Economy of the United States: Readings in Military Industry and Economy*. New York: St. Martin's Press.

Pettini, M. 2007. *Geometry and Topology in Hamiltonian Dynamics*. New York: Springer.

Protter, P. 1990. *Stochastic Integration and Differential Equations*. New York: Springer.

Ullmann, J.E. 1988. *The Anatomy of Industrial Decline*. Westport, CT: Greenwood-Quorum.

Wallace, D. 2011. Discriminatory mass de-housing and low-weight births: Scales of geography, time, and level. *Journal of Urban Health* 88: 454–468.

Wallace, R. 2013. A new formal approach to evolutionary processes in socioeconomic systems. *Journal of Evolutionary Economics* 23: 1–15.

Wallace, R. 2014. A new formal perspective on 'Cambrian explosions'. *Comptes Rendus Biologies* 337: 1–5.

Wallace, R. 2015. *An Ecosystem Approach to Economic Stabilization: Escaping the Neoliberal Wilderness*. Routledge Advances in Heterodox Economics. New York: Routledge.

Wallace, D., and R. Wallace. 1998. *A Plague on Your Houses*. New York: Verso.

Wallace, R., and D. Wallace. 2017. *Gene Expression and Its Discontents: The Social Production of Chronic Disease*, 2nd ed. New York: Springer.

Chapter 11
Social Psychopathology: Military Doctrine and the Madness of Crowds

Summary Institutions, commercial enterprises, communities, political entities, and the social milieus in which they are enmeshed must engage in cognitive process to address day-to-day contingencies and incorporate the learning needed to successfully adapt to larger evolutionary selection pressures. Failure of cognition, we show using control and information theory models, will be driven by environmental challenges that can force such systems into persistent ground states where cognitive process becomes pathologically fixated. The consequent path-dependent developmental path then remains pathological in the absence of sufficiently sustained external correction pressures. Under combat conditions this might translate into "all possible targets are enemies." Among human tribal groups, "only we are real people," and so on. These powerful dynamics preclude the engineering resilience approach of current US security doctrine: While it may be possible to ensure return to normal function for relatively simple power and communications networks under moderate perturbation, extension of the idea to socioeconomic entities is a ludicrous fantasy. Ecosystem and evolutionary perspectives that recognize the possibility of path-dependence and long-term eutrophication, in various forms, are more relevant, and may lead to realistic and sustainable policy objectives.

11.1 Introduction

> The term "resilience" [as used in US security doctrine] refers to the ability to adapt to changing conditions and withstand and rapidly recover from disruption due to emergencies... The United States officially recognized resilience in national doctrine in the 2010 National Security Strategy, which states that we must enhance our resilience – the ability to adapt to changing conditions and prepare for, withstand, and rapidly recover from disruption. (US Dept. of Homeland Security 2016).

> Translated into military terms Bergson's "elan vital" [the all-conquering will] became the doctrine of the offensive. In proportion as a defensive strategy gave way to an offensive strategy, the attention paid to the Belgian frontier gradually gave way in favor of a progressive shift of gravity eastward toward the point where a French offensive could be launched to break through to the Rhine.(Tuchman, The Guns of August, 1962).

Western cultural atomism, it has been argued (Wallace 2015a,b Chap. 1, and the references therein), limits, and indeed badly deforms, theory in economics, evolution, and human psychology. An example of ideological deformation in

R. Wallace, *Computational Psychiatry*, DOI 10.1007/978-3-319-53910-2_11

"resilience" research is the work of Gao et al. (2016) published by the prestigious journal Nature—and supported by the Army Research Laboratories and the Defense Threat Reduction Agency—that seeks "universal resilience patterns in complex networks." Gao et al. (2016) claim

> [Our] analytical results unveil the network characteristics that can enhance or diminish resilience, offering ways to prevent the collapse of ecological, biological or economic systems, and guiding the design of technological systems resilient to both internal failures and environmental changes.

This is a predetermined result clearly driven by the doctrine quoted above. Military-funded research is centrally tasked with implementing doctrine, here constrained to deliver a simplistic engineering resilience involving the ability to bounce back to near normal function after significant perturbation. US security doctrine thus inhibits exploration more complex—and far more likely—scenarios that do not have politically palatable outcomes.

This is indeed reminiscent of "elan vital", a doctrine that involved, in addition to command blindness regarding German incursion through Belgium, such tactics as mass charges into concentrations of machine guns entrenched behind barbed wire. Today, US security doctrine calls for "Resilience." A 100 years has not been enough time for Western military practitioners to appreciate the deadly burdens of tactical and strategic fantasy. The recent US occupation of Iraq also comes to mind.

Apparently this is all of a piece.

We know that enterprises and institutions—cognitive entities that can learn from experience and incorporate that learning into corporate culture—are subject to evolutionary selection pressures strongly enforcing path dependence (Hodgso and Knudsen 2010; Wallace 2015a, and the references therein). "Buggy whip" industries become extinct in the face of significant market shifts if they do not adapt rapidly enough.

Evolutionary dynamics profoundly challenge the engineering resilience requirements of US defense and homeland security doctrine for enterprises, institutions, and socioeconomic systems. Intractable difficulties emerge from the powerful cognitive processes that must incorporate learning into corporate culture for successful adaptation, while crafting responses under shifting day-to-day market demands. Most particularly, facing environmental pressures, in a large sense, cognitive entities for which information about the real world is an essential control signal can be rapidly driven into a pathological ground state of policy paralysis and fixation that may ensure their ultimate demise. We will provide models of this taken from control and information theories.

In sum, the cognitive dynamics essential to short-term institutional function and long-term adaptation are synergistic with path-dependent evolutionary process in precluding doctrine and policy based on ideas of ensuring simple bounce back to normal following serious disturbance.

11.2 A Control Theory Model

Civilian aircraft can, often for considerable periods, remain in stable flight as long as their center of pressure is well behind their center of gravity: Perturbations of trajectory quickly diminish. By contrast, cognitive systems like human institutions usually operate on rapidly shifting topological roadways of complicated "real time" multimodal demand. These are highly turbulent topologies, and, to function successfully, cognitive systems embedded in them must receive constant flows of sufficiently detailed descriptive information. One analog is that of driving at high speed on a twisting, pot-holed road at night, a feat demanding not only a reliable vehicle and fast reflexes, but also excellent headlights. See Chap. 8.

We thus argue that a cognitive process operating under such constraints is inherently unstable in the sense of control theory's Data Rate Theorem, viewing incoming information about the rapidly shifting topology of demand as the control signal. This fact has profound implications, particularly for path-dependent phenomena.

The canonical first model uses a linear expansion around a nonequilibrium steady state for which the rate of control information is sufficient for stability.

Paraphrasing Nair et al. (2007), the Data Rate Theorem (DRT) contends that there is a minimum rate at which control information must be provided for an inherently unstable system to remain stable. Making a linear expansion near a nonequilibrium steady state, an n-dimensional vector of system parameters at time t, x_t, determines the state at time $t+1$ according to Fig. 11.1, with

$$x_{t+1} = \mathbf{A}x_t + \mathbf{B}u_t + W_t \tag{11.1}$$

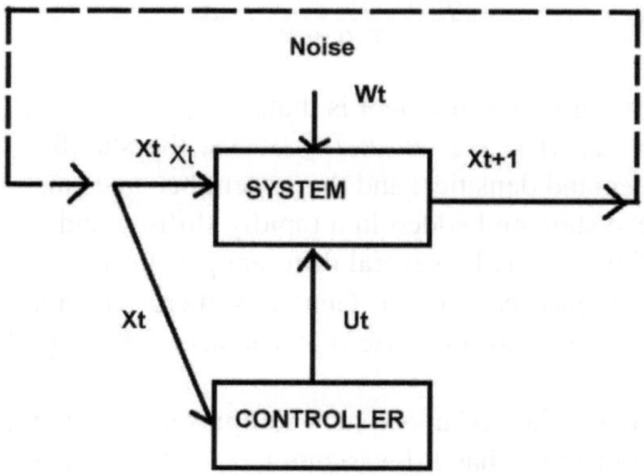

Fig. 11.1 A linear expansion near a nonequilibrium steady state for an inherently unstable control system. **A** and **B** are square matrices, x_t the vector of system parameters at time t, u_t the vector of control signals at time t, W_t a white noise vector, and $x_{t+1} = \mathbf{A}x_t + \mathbf{B}u_t + W_t$. The Data Rate Theorem states that the minimum rate at which control information must be provided for system stability is $\mathcal{H} > \log[|\det[\mathbf{A}^m|]$, where $\mathbf{A}^m$ is the subcomponent of **A** having eigenvalues ≥ 1

$\mathbf{A}, \mathbf{B}$ are fixed $n \times n$ matrices, u_t is the vector of control information, and W_t is an n-dimensional vector of white noise. The DRT under such a condition states that the minimum control information rate for stability, $\mathcal{H}$, satisfies the inequality

$$\mathcal{H} > \log[|\det(\mathbf{A}^m)|] \equiv a_0 \quad (11.2)$$

where, for $m \leq n$, $\mathbf{A}^m$ is the subcomponent of $\mathbf{A}$ having eigenvalues ≥ 1. The right-hand side of Eq. (11.2) defines the rate at which the system generates topological information. A more general version of this result is derived in the Mathematical Appendix using a Rate Distortion Theorem approach.

Defining a parameter ρ as a scalar representing the magnitude of some index of multimodal system demand that will be explored further below, we can extend Eq. (11.2) as

$$\mathcal{H}(\rho) > f(\rho)a_0 \quad (11.3)$$

where a_0 is an inherent system parameter and $f(\rho)$ is a positive, monotonically increasing function of environmental stress. The Mathematical Appendix uses a Black–Scholes model to approximate the "cost" of $\mathcal{H}$ as a function of the "investment" ρ. The first approximation is linear, so that $\mathcal{H} \approx \kappa_1 \rho + \kappa_2$.

Expanding $f(\rho)$ to similar order, i.e.,

$$f(\rho) \approx \kappa_3 \rho + \kappa_4 \quad (11.4)$$

defines the limit condition for stability as

$$\mathcal{T} \equiv \frac{\kappa_1 \rho + \kappa_2}{\kappa_3 \rho + \kappa_4} > a_0 \quad (11.5)$$

For $\rho = 0$, the stability requirement is that $\kappa_2/\kappa_4 > a_0$. At large ρ the condition becomes $\kappa_1/\kappa_2 > a_0$. If $\kappa_2/\kappa_4 \gg \kappa_1/\kappa_2 < a_0$, the stability condition may be breached at high demand densities, and the system becomes instable. See Fig. 11.2.

For a cognitive system embedded in a rapidly shifting and complicated demand stream "highway," there may be several different ρ-values that are not independent but interact and influence each other. Generally there will not be a simple scalar index of demand, but an unsymmetric $n \times n$ matrix $\hat{\rho}$ having elements $\rho_{i,j}, i,j = 1 \ldots n$.

Can we still find a scalar "ρ" under such complex circumstances so that Fig. 11.2 applies? An $n \times n$ matrix $\hat{\rho}$ has n invariants $r_i, i = 1 \ldots n$ that remain fixed when a certain class of "principal component analysis" transformations is applied. The invariants are found via the *characteristic equation* of the matrix

$$\mathcal{P}(\lambda) = \det(\hat{\rho} - \lambda I_n) = \lambda^n + r_1 \lambda^{n-1} + \cdots + r_{n-1}\lambda + r_n \quad (11.6)$$

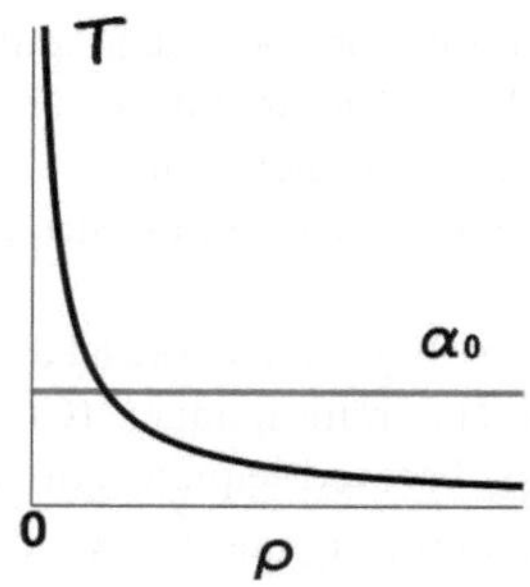

Fig. 11.2 The *horizontal line* represents the critical limit a_0. If $a_0 < \kappa_2/\kappa_4 \gg \kappa_1/\kappa_3 < a_0$, at some intermediate value of demand ρ, the temperature analog $\mathcal{T} \equiv (\kappa_1\rho + \kappa_2)/(\kappa_3\rho + \kappa_4)$ falls below that limit, the system becomes "supercooled," and failure becomes increasingly probable

det is the determinant, λ is a parameter that is an element of a ring, and I_n the $n \times n$ identity matrix. The invariants are the coefficients of λ in the polynomial $\mathcal{P}(\lambda)$, normalized so that the coefficient of λ^n is 1.

The r_i, we claim, can be used to construct an invariant scalar measure, subject to the approximation inherent in projecting an $n \times n$ matrix onto one dimension.

Since $n \times n$ matrices themselves form a ring,

$$\mathcal{P}(\hat{\rho}) = 0 \times I_n \tag{11.7}$$

and a matrix satisfies its own characteristic equation.

An $n \times n$ matrix will thus have n such invariants from which a new scalar index $\Gamma = g(r_1, \ldots, r_n)$ can be constructed to replace ρ in Eq. (11.5). The first invariant will be the trace and the last $\pm$ the determinant of $\hat{\rho}$.

We reiterate that reduction of the complicated interaction matrix $\hat{\rho}$ to the scalar Γ is an ambiguous matter, at best an approximation that must be dynamically fitted to each fog-of-war ecosystem. There is, as a consequence, no one-size-fits-all simplification methodology, although there may be equivalence classes of different systems that can be mapped onto a particular method.

This development is another example of the "Rate Distortion Manifold" (RDM) of Glazebrook and Wallace (2009), or the equivalent "Generalized Retina" of Wallace and Wallace (2013, Sect. 10.1) in which high dimensional data flows can be projected down onto lower dimensional, shifting and tunable "tangent spaces" with minimal loss of essential information.

The RDM approach is in contrast to the treatment of engineering resilience—return to a single stable mode after perturbation—by Gao et al. (2016) who condense network topologies by defining an "effective state" of a multidimensional network system using the average nearest-neighbor activity. They thus collapse a multidimensional dynamic equation into one-dimensional form. By contrast, here the DRT provides an inherently one-dimensional format whose complexities reside in the matrix $\hat{\rho}$, leading to an understanding of complex ecological resilience in the sense of Holling (1973), that is, the existence of a number of quasi-stable modes, with the possibility of transitions between them driven by changes in context indexed by an appropriately chosen "temperature" analog $\mathcal{T}$, defined now in terms of the complicated "retina" index Γ rather than ρ.

$\mathcal{T}$ will, we can show, undergo dynamics strongly affected by embedding stochastic circumstances.

How is a control signal u_t in Fig. 11.1 expressed in the system response x_{t+1}? In standard Rate Distortion Theorem manner (Cover and Thomas 2006), we deterministically retranslate an observed sequence of system outputs $X^i = x_1^i, x_2^i, \ldots$ into a sequence of possible control signals $\hat{U}^i = \hat{u}_0^i, \hat{u}_1^i, \ldots$ and to compare that sequence with the original control sequence $U^i = u_0^i, u_1^i, \ldots$, with the difference between them having a particular value under some chosen distortion measure and hence having an average distortion

$$< d >= \sum_i p(U^i) d(U^i, \hat{U}^i) \tag{11.8}$$

$p(U^i)$ is the probability of the sequence U^i. $d(U^i, \hat{U}^i)$ is the distortion between U^i and the sequence of control signals that has been deterministically reconstructed from system output.

It is then possible to apply a Rate Distortion argument. The Rate Distortion Theorem asserts that there is a Rate Distortion Function that determines the minimum channel capacity—$R(D)$—necessary to keep the average distortion below some fixed limit D. Again, see Cover and Thomas (2006) for details. Taking Feynman's (2000) interpretation of information as a form of free energy, it is possible to construct a Boltzmann-like pseudoprobability in the temperature analog $\mathcal{T}$:

$$dP(R, \mathcal{T}) = \frac{\exp[-R/\mathcal{T}]dR}{\int_0^\infty \exp[-R/\mathcal{T}]dR} \tag{11.9}$$

since higher $\mathcal{T}$ must necessarily be associated with greater channel capacity.

The integral of the denominator is interpreted as a statistical mechanical partition function, allowing definition of a "second order" free energy Morse Function F (Pettini 2007):

$$\exp[-F/\mathcal{T}] = \int_0^\infty \exp[-R/\mathcal{T}]dR = \mathcal{T} \tag{11.10}$$

This gives $F(\mathcal{T}) = -\mathcal{T} \log[\mathcal{T}]$.

Similarly, a "second order" entropy-analog can also be defined:

$$\mathcal{S} \equiv F(\mathcal{T}) - \mathcal{T} dF/d\mathcal{T} = \mathcal{T} \tag{11.11}$$

An analog to the Onsager treatment of nonequilibrium thermodynamics (de Groot and Mazur 1984) can now be used, using the gradient of $\mathcal{S}$ in $\mathcal{T}$ as the driving factor, and a stochastic version of the standard Onsager equation emerges:

$$dT_t \approx (\mu dS/dT)dt + \beta T_t dW_t$$
$$= \mu dt + \beta T_t dW_t \tag{11.12}$$

μ is an analog to a diffusion coefficient and β is the magnitude of the impinging white noise dW_t.

By a now-familiar argument, applying the Ito chain rule to $\log(\mathcal{T})$ in Eq. (11.12) (Protter 1990), using Jensen's inequality for a concave function gives the nonequilibrium steady state expectation of $\mathcal{T}$ as

$$E(\mathcal{T}_t) \geq \frac{\mu}{\beta^2/2} \tag{11.13}$$

μ parameterizes the efforts of the control apparatus to maintain stability. Thus, according to this approximation, rising system noise—β—can significantly increase the probability that $\mathcal{T}$ falls below the critical limit, triggering a failure of control.

It is critical to note that, since $E(\mathcal{T})$ is an expectation, according to the model, there will always be some probability that $\mathcal{T}$ will fall below the critical value a_0 in the multimodal expression

$$\mathcal{T} \approx \frac{\kappa_1 \Gamma + \kappa_2}{\kappa_3 \Gamma + \kappa_4} > a_0 \tag{11.14}$$

Raising μ and limiting β decreases that probability, but does not eliminate it. Instability remains possible at every level of μ.

β itself can, of course, have further structure, leading to more complicated dynamics.

11.3 A Cognitive Model

The DRT argument above indicates a significantly raised probability of a transition between stable and unstable behavior if the temperature analog $\mathcal{T}$ from Eq. (11.14) falls below a critical value. It is possible to extend the argument, involving more complicated patterns of phase transition, using the cognitive approach of Atlan and Cohen (1998). They define a system as cognitive if it compares incoming signals with a learned or inherited picture of the world, and then actively chooses a response from a larger set of what is possible to it. Choice implies the existence of—and indeed virtually defines—an information source, since it reduces uncertainty in a formal way (Wallace 2012, 2015a,b).

Assuming such a "dual" information source is associated with the inherently unstable cognitive system under study, an equivalence class algebra can be constructed by choosing different system origin states and defining the equivalence of subsequent states at a later time by the existence of a high probability path connecting them to the same origin state. Disjoint partition by equivalence class is analogous to orbit equivalence classes in dynamical systems. This inherently defines a symmetry groupoid associated with the cognitive process (Wallace 2012). Groupoids are generalizations of group symmetries in which there is not necessarily a product defined for each possible element pair (Weinstein 1996). The canonical simple example is the disjoint union of different groups.

The equivalence classes across possible origin states then define a set of information sources dual to different cognitive states available to the inherently unstable cognitive system. These instantiate a very large groupoid, with each orbit corresponding to a transitive groupoid whose disjoint union is the full groupoid. Each subgroupoid is associated with a dual information source, and larger groupoids will have richer dual information sources than smaller.

Take X_{G_i} as the dual information source associated with some groupoid element G_i. Assuming Eqs. (11.5)–(11.7), it becomes possible to construct another Morse Function (Pettini 2007), again using Eq. (11.14) to define the "temperature" $\mathcal{T}$.

Let $H(X_{G_i}) \equiv H_{G_i}$ be the Shannon uncertainty of the information source associated with the groupoid element G_i. Define a Boltzmann-like pseudoprobability:

$$P[H_{G_i}] \equiv \frac{\exp[-H_{G_i}/\mathcal{T}]}{\sum_j \exp[-H_{G_j}/\mathcal{T}]} \tag{11.15}$$

The sum is now over the different possible cognitive modes of the full system.
A "free energy" Morse Function $\mathcal{F}$ can again be defined:

$$\exp[-\mathcal{F}/\mathcal{T}] \equiv \sum_j \exp[-H_{G_j}/\mathcal{T}]$$

$$\mathcal{F} = -\mathcal{T} \log\left[\sum_j \exp[-H_{G_j}/\mathcal{T}]\right] \tag{11.16}$$

Using the underlying groupoid generalized symmetries, it is possible to invoke an obvious extension of Landau's version of phase transition (Pettini 2007). Landau argued that spontaneous symmetry breaking of a group structure represents phase change in physical systems, with higher energies available at higher temperatures being more symmetric. The shift between symmetries is then punctuated in the temperature index, here the $\mathcal{T}$ constructed in terms of the scalar index $\Gamma = g(r_1, \ldots, r_n)$. This is now in the context of groupoid rather than group symmetries. Typically, for physical systems, there are only a few phases possible, with sharply punctuated transitions between them as $\mathcal{T}$ decreases.

Sufficient conditions for the pathological stability of the hypercondensed "ground state" cognitive phase in which "all possible targets are enemies," or some analog, can be explored using the approach of Wallace (2016).

Assuming a vector of parameters $\mathbf{J}$ measuring deviations from a pathological nonequilibrium steady state, the "free energy" analog $\mathcal{F}$ in Eq. (11.16) can be used to define a new "entropy" scalar as the Legendre transform

$$\mathcal{S} \equiv \mathcal{F}(\mathbf{J}) - \mathbf{J} \cdot \nabla_{\mathbf{J}} \mathcal{F} \tag{11.17}$$

Again, a first order dynamic equation follows using the stochastic version of the Onsager formalism from nonequilibrium thermodynamics (de Groot and Mazur 1984)

$$dJ_t^i \approx \Big(\sum_k \mu_{i,k} \partial S / \partial J_t^k\Big) dt + \sigma_i J_t^i dB_t \tag{11.18}$$

$\mu_{i,k}$ defines a diffusion matrix, the σ_i are parameters, and dB_t represents a noise that may not the usual Brownian motion under undifferentiated white noise.

If it is possible to factor out J^i, then Eq. (11.18) can be represented as:

$$dJ_t^i = J_t^i dY_t^i \tag{11.19}$$

Here, Y_t^i may be a complicated stochastic process.

Equation (11.20) can now be solved in expectation using the Doleans-Dade exponential (Protter 1990):

$$E(J_t^i) \propto \exp(Y_t^i - 1/2[Y_t^i, Y_t^i]) \tag{11.20}$$

where $[Y_t^i, Y_t^i]$ is the quadratic variation of the stochastic process Y_t^i (Protter 1990). Using the Mean Value Theorem, if

$$1/2d[Y_t^i, Y_t^i]/dt > dY_t^i/dt \tag{11.21}$$

then the pathological ground state is stable. That is, deviations from nonequilibrium steady state measured by J_t^i then converge in expectation to 0. Thus sufficient ongoing fog-of-war noise or other environmental stress which determines the quadratic variation terms can lock-in the failure of "target discrimination," in a large sense. There will be no return to the "normal" state under such circumstances: the "engineering resilience" of Gao et al. (2016), and of current US security doctrine, fails.

As described above, similar dynamics arise in ecosystem resilience theory (Holling 1973) that characterizes multiple quasi-stable nonequilibrium steady states among interacting populations. The standard example is that pristine alpine lake ecosystems, which have limited nutrient inflows, can be permanently shifted into a toxic eutrophic state by excess nutrient influx—a sewage leak, fertilizer runoff, and so on. Once shifted, the lake ecology will remain trapped in a mode of recurrent "red tide"-like plankton blooms even after sewage or fertilizer inflow is stemmed.

The quadratic variation can be estimated from time series data using the spectral methods of Dzhaparidze and Spreij (1994), taken from the literature on financial engineering, in which these approaches are standard.

11.4 Discussion and Conclusions

Institutions, commercial enterprises, communities, and their enmeshing social structures must engage in cognitive process to address rapidly changing patterns of challenge and opportunity, and, more slowly, incorporate the learning necessary for successful adaptation to shifts in large-scale evolutionary selection pressures. Failure of cognition, we have indicated, can be triggered by environmental challenges that drive such structures into highly persistent ground states where cognitive process becomes pathologically fixated, initiating a pathological developmental pathway.

Under combat conditions this dynamic might translate into "all possible targets are enemies."

Among human tribal groups of various sorts, something like "only we are real people" emerges, leading to or enforcing social fragmentation.

Failing enterprises often focus on analogs to counting boxes of office pens and numbers of headquarters toilet paper rolls instead of on new products or market strategies.

As described above, the accelerating political and public health instabilities following the Cold War-induced deindustrialization of the USA provide a powerful case history (Wallace and Wallace 2010, Chap. 7; Wallace 2015a, Chap. 7). Details of how US deindustrialization was driven by 50 years of Cold War can be found in Ullmann (1988), Melman (1971), and related works.

Wallace and Wallace (1998) and Wallace (2011) explore the long-term impacts of a "planned shrinkage" policy aimed against minority voting blocks in New York City and implemented by the targeted withdrawal of fire extinguishment resources from poverty-stricken, high fire incidence, high population density neighborhoods. This deliberate policy triggered rapid downward spirals of social and physical disintegration causing great exacerbation of multiple indices of morbidity, mortality, and criminal activity.

Readers will have their own examples.

None of this resembles the engineering resilience of current US security doctrine as expressed in the opening quotation and instantiated by Gao et al. (2016). While it may be possible to stabilize relatively simple power and communications networks under moderate perturbation, extension of the approach to institutions, enterprises, economic structures, and communities is ludicrous. Ecosystem and institutional evolutionary perspectives that recognize the possibility of persistent eutrophication, in various forms, are more to the point (e.g., Holling 1973), although the implications will most surely not please the present US security establishment, whose many wishful-thinking "Elan"-like doctrines, and the policies based on them, have contributed materially to the nation's decline (Wallace 2015a, Chap. 7; Wallace and Wallace 2010, Chap. 7, and the references therein).

Ecosystem resilience and institutional evolutionary theories may, if properly adapted, provide tools actually useful in achieving realistic policy goals.

References

Atlan, H., and I. Cohen. 1998. Immune information, self-organization, and meaning. *International Immunology* 10: 711–717.

Cover, T., and J. Thomas. 2006. *Elements of Information Theory*, 2nd ed. New York: Wiley.

de Groot, S., and P. Mazur. 1984. *Nonequilibrium Thermodynamics*. New York: Dover.

Dzhaparidze, K., and P. Spreij. 1994. Spectral characterization of the optional quadratic variation process. *Stochastic Processes and Their Applications* 54: 165–174.

Feynman, R. 2000. *Lectures on Computation*. New York: Westview Press.

Gao, H., B. Barzel, and A. Barabasi. 2016. Universal resilience patterns in complex networks. *Nature* 530: 307–312. See erratum. doi: 10.1038/nature18019.

Glazebrook, J., and R. Wallace. 2009. Rate distortion manifolds as model spaces for cognitive information. *Informatica* 33: 309–345.

Hodgson, G., and T. Knudsen. 2010. *Darwin's Conjecture: The Search for General Principles of Social and Economic Evolution*. Chicago, IL: University of Chicago Press.

Holling, C. 1973. Resilience and stability of ecological systems. *Annual Reviews of Ecological Systematics* 4: 1–23.

Melman, S. 1971. *The War Economy of the United States: Readings in Military Industry and Economy*. New York: St. Martin's Press.

Nair, G., F. Fagnani, S. Zampieri, and R. Evans. 2007. Feedback control under data rate constraints: An overview. *Proceedings of the IEEE* 95: 108–137.

Pettini, M. 2007. *Geometry and Topology in Hamiltonian Dynamics*. New York: Springer.

Protter, P. 1990. *Stochastic Integration and Differential Equations*. New York: Springer.

Tuchman, B. 1962. *The Guns of August*. New York: Dell.

Ullmann, J.E. 1988. *The Anatomy of Industrial Decline*. Westport, CT: Greenwood-Quorum.

US Dept. of Homeland Security. 2016. https://www.dhs.gov/topic/resilience.

Wallace, D. 2011. Discriminatory mass de-housing and low-weight births: Scales of geography, time, and level. *Journal of Urban Health* 88: 454–468.

Wallace, R. 2012. Consciousness, crosstalk and the mereological fallacy: An evolutionary perspective. *Physics of Life Reviews* 9: 426–453.

Wallace, R. 2015a. *An Ecosystem Approach to Economic Stabilization: Escaping the Neoliberal Wilderness*. New York: Routledge Advances in Heterodox Economics.

Wallace, R. 2015b. *An Information Approach to Mitochondrial Dysfunction: Extending Swerdlow's Hypothesis*. Singapore: World Scientific.

Wallace, R. 2016. Subtle noise structures as control signals in high-order biocognition. *Physics Letters A* 380: 726–729.

Wallace, D., and R. Wallace. 1998. *A Plague on Your Houses*. New York: Verso.

Wallace, R., and D. Wallace. 2010. *Gene Expression and Its Discontents: The Social Production of Chronic Disease*. New York: Springer.

Wallace, R., and D. Wallace. 2013. *A Mathematical Approach to Multilevel, Multiscale Health Interventions: Pharmaceutical Industry Decline and Policy Response*. London: Imperial College Press.

Weinstein, A. 1996. Groupoids: Unifying internal and external symmetry. *Notices of the American Mathematical Society* 43: 744–752.

Chapter 12
Mathematical Appendix

12.1 Groupoids

Following Weinstein (1996) closely, a groupoid, G, is defined by a base set A upon which some mapping—a morphism—can be defined. Note that not all possible pairs of states (a_j, a_k) in the base set A can be connected by such a morphism. Those that can define the groupoid element, a morphism $g = (a_j, a_k)$ having the natural inverse $g^{-1} = (a_k, a_j)$. Given such a pairing, it is possible to define "natural" end-point maps $\alpha(g) = a_j, \beta(g) = a_k$ from the set of morphisms G into A, and a formally associative product in the groupoid $g_1 g_2$ provided $\alpha(g_1 g_2) = \alpha(g_1), \beta(g_1 g_2) = \beta(g_2)$, and $\beta(g_1) = \alpha(g_2)$. Then the product is defined, and associative, $(g_1 g_2)g_3 = g_1(g_2 g_3)$.

In addition, there are natural left and right identity elements λ_g, ρ_g such that $\lambda_g g = g = g\rho_g$ (Weinstein 1996).

An orbit of the groupoid G over A is an equivalence class for the relation $a_j \sim Ga_k$ if and only if there is a groupoid element g with $\alpha(g) = a_j$ and $\beta(g) = a_k$. Following Cannas DaSilva and Weinstein (1999), we note that a groupoid is called transitive if it has just one orbit. The transitive groupoids are the building blocks of groupoids in that there is a natural decomposition of the base space of a general groupoid into orbits. Over each orbit there is a transitive groupoid, and the disjoint union of these transitive groupoids is the original groupoid. Conversely, the disjoint union of groupoids is itself a groupoid.

The isotropy group of $a \in X$ consists of those g in G with $\alpha(g) = a = \beta(g)$. These groups prove fundamental to classifying groupoids.

If G is any groupoid over A, the map $(\alpha, \beta) : G \to A \times A$ is a morphism from G to the pair groupoid of A. The image of (α, β) is the orbit equivalence relation $\sim G$, and the functional kernel is the union of the isotropy groups. If $f : X \to Y$ is a function, then the kernel of f, $\ker(f) = [(x_1, x_2) \in X \times X : f(x_1) = f(x_2)]$ defines an equivalence relation.

Groupoids may have additional structure. As Weinstein (1996) explains, a groupoid G is a topological groupoid over a base space X if G and X are topological spaces and α, β and multiplication are continuous maps. A criticism sometimes

R. Wallace, *Computational Psychiatry*, DOI 10.1007/978-3-319-53910-2_12

applied to groupoid theory is that their classification up to isomorphism is nothing other than the classification of equivalence relations via the orbit equivalence relation and groups via the isotropy groups. The imposition of a compatible topological structure produces a nontrivial interaction between the two structures. In Sect. 2.7 we introduced a metric structure on manifolds of related information sources, producing such interaction.

In essence, a groupoid is a category in which all morphisms have an inverse, here defined in terms of connection to a base point by a meaningful path of an information source dual to a cognitive process.

As Weinstein (1996) points out, the morphism (α, β) suggests another way of looking at groupoids. A groupoid over A identifies not only which elements of A are equivalent to one another (isomorphic), but *it also parameterizes the different ways (isomorphisms) in which two elements can be equivalent*, i.e., in our context, all possible information sources dual to some cognitive process. Given the information-theoretic characterization of cognition presented above, this produces a full modular cognitive network in a highly natural manner.

Brown (1987) describes the fundamental structure as follows:

> A groupoid should be thought of as a group with many objects, or with many identities... A groupoid with one object is essentially just a group. So the notion of groupoid is an extension of that of groups. It gives an additional convenience, flexibility and range of applications...
>
> EXAMPLE 1. A disjoint union [of groups] $G = \cup_\lambda G_\lambda, \lambda \in \Lambda$, is a groupoid: the product ab is defined if and only if a, b belong to the same G_λ, and ab is then just the product in the group G_λ. There is an identity 1_λ for each $\lambda \in \Lambda$. The maps α, β coincide and map G_λ to $\lambda, \lambda \in \Lambda$.
>
> EXAMPLE 2. An equivalence relation R on [a set] X becomes a groupoid with $\alpha, \beta : R \rightarrow X$ the two projections, and product $(x, y)(y, z) = (x, z)$ whenever $(x, y), (y, z) \in R$. There is an identity, namely (x, x), for each $x \in X$.

Weinstein (1996) makes the following fundamental point:

> Almost every interesting equivalence relation on a space B arises in a natural way as the orbit equivalence relation of some groupoid G over B. Instead of dealing directly with the orbit space B/G as an object in the category S_{map} of sets and mappings, one should consider instead the groupoid G itself as an object in the category G_{htp} of groupoids and homotopy classes of morphisms.

The groupoid approach has become quite popular in the study of networks of coupled dynamical systems which can be defined by differential equation models (e.g., Golubitsky and Stewart 2006).

12.2 The Tuning Theorem

Messages from an information source, seen as symbols x_j from some alphabet, each having probabilities P_j associated with a random variable X, are "encoded" into the language of a "transmission channel," a random variable Y with symbols y_k, having probabilities P_k, possibly with error. Someone receiving the symbol y_k then retranslates it (without error) into some x_k, which may or may not be the same as the x_j that was sent.

More formally, the message sent along the channel is characterized by a random variable X having the distribution

$$P(X = x_j) = P_j, j = 1, \ldots, M.$$

The channel through which the message is sent is characterized by a second random variable Y having the distribution

$$P(Y = y_k) = P_k, k = 1, \ldots, L.$$

Let the joint probability distribution of X and Y be defined as

$$P(X = x_j, Y = y_k) = P(x_j, y_k) = P_{j,k}$$

and the conditional probability of Y given X as

$$P(Y = y_k | X = x_j) = P(y_k | x_j).$$

Then the Shannon uncertainty of X and Y independently and the joint uncertainty of X and Y together are defined, respectively, as

$$H(X) = -\sum_{j=1}^{M} P_j \log(P_j)$$

$$H(Y) = -\sum_{k=1}^{L} P_k \log(P_k)$$

$$H(X, Y) = -\sum_{j=1}^{M} \sum_{k=1}^{L} P_{j,k} \log(P_{j,k}) \quad (12.1)$$

The *conditional uncertainty* of Y given X is defined as

$$H(Y|X) = -\sum_{j=1}^{M} \sum_{k=1}^{L} P_{j,k} \log[P(y_k|x_j)] \quad (12.2)$$

For any two stochastic variates X and Y, $H(Y) \geq H(Y|X)$, as knowledge of X generally gives some knowledge of Y. Equality occurs only in the case of stochastic independence.

Since $P(x_j, y_k) = P(x_j)P(y_k|x_j)$, then $H(X|Y) = H(X, Y) - H(Y)$.

The information transmitted by translating the variable X into the channel transmission variable Y—possibly with error—and then retranslating without error the transmitted Y back into X is defined as

$$I(X|Y) \equiv H(X) - H(X|Y) = H(X) + H(Y) - H(X,Y) \tag{12.3}$$

See Cover and Thomas (2006) for details. The essential point is that if there is no uncertainty in X given the channel Y, then there is no loss of information through transmission. In general this will not be true, and herein lies the essence of the theory.

Given a fixed vocabulary for the transmitted variable X, and a fixed vocabulary and probability distribution for the channel Y, we may vary the probability distribution of X in such a way as to maximize the information sent. The capacity of the channel is defined as

$$C \equiv \max_{P(X)} I(X|Y) \tag{12.4}$$

subject to the subsidiary condition that $\sum P(X) = 1$.

The critical trick of the Shannon Coding Theorem for sending a message with arbitrarily small error along the channel Y at any rate $R < C$ is to encode it in longer and longer "typical" sequences of the variable X; that is, those sequences whose distribution of symbols approximates the probability distribution $P(X)$ above which maximizes C.

If $S(n)$ is the number of such "typical" sequences of length n, then

$$\log[S(n)] \approx nH(X)$$

where $H(X)$ is the uncertainty of the stochastic variable defined above. Some consideration shows that $S(n)$ is much less than the total number of possible messages of length n. Thus, as $n \to \infty$, only a vanishingly small fraction of all possible messages is meaningful in this sense. This observation, after some considerable development, is what allows the Coding Theorem to work so well. In sum, the prescription is to encode messages in typical sequences, which are sent at very nearly the capacity of the channel. As the encoded messages become longer and longer, their maximum possible rate of transmission without error approaches channel capacity as a limit. Again, the standard references provide details.

This approach can be, in a sense, inverted to give a "tuning theorem" variant of the coding theorem.

Telephone lines, optical wave guides, and the tenuous plasma through which a planetary probe transmits data to earth may all be viewed in traditional information-theoretic terms as a *noisy channel* around which we must structure a message so as to attain an optimal error-free transmission rate.

Telephone lines, wave guides, and interplanetary plasmas are, relatively speaking, fixed on the timescale of most messages, as are most sociogeographic networks.

Indeed, the capacity of a channel is defined by varying the probability distribution of the "message" process X so as to maximize $I(X|Y)$.

Suppose there is some message X so critical that its probability distribution must remain fixed. The trick is to fix the distribution $P(x)$ but *modify the channel*—i.e., tune it—so as to maximize $I(X|Y)$. The *dual* channel capacity C^* can be defined as

$$C^* \equiv \max_{P(Y),P(Y|X)} I(X|Y) \tag{12.5}$$

But

$$C^* = \max_{P(Y),P(Y|X)} I(Y|X)$$

since

$$I(X|Y) = H(X) + H(Y) - H(X,Y) = I(Y|X).$$

Thus, in a purely formal mathematical sense, *the message transmits the channel*, and there will indeed be, according to the Coding Theorem, a channel distribution $P(Y)$ which maximizes C^*.

One may do better than this, however, by modifying the channel matrix $P(Y|X)$. Since

$$P(y_j) = \sum_{i=1}^{M} P(x_i)P(y_j|x_i),$$

$P(Y)$ is entirely defined by the channel matrix $P(Y|X)$ for fixed $P(X)$ and

$$C^* = \max_{P(Y),P(Y|X)} I(Y|X) = \max_{P(Y|X)} I(Y|X).$$

Calculating C^* requires maximizing the complicated expression

$$I(X|Y) = H(X) + H(Y) - H(X,Y)$$

which contains products of terms and their logs, subject to constraints that the sums of probabilities are 1 and each probability is itself between 0 and 1. Maximization is done by varying the channel matrix terms $P(y_j|x_i)$ within the constraints. This is a difficult problem in nonlinear optimization. However, for the special case $M = L$, C^* may be found by inspection.

If $M = L$, then choose

$$P(y_j|x_i) = \delta_{j,i}$$

where $\delta_{i,j}$ is 1 if $i = j$ and 0 otherwise. For this special case

$$C^* \equiv H(X)$$

with $P(y_k) = P(x_k)$ for all k. *Information is thus transmitted without error when the channel becomes "typical" with respect to the fixed message distribution* $P(X)$.

If $M < L$ matters reduce to this case, but for $L < M$ information must be lost, leading to Rate Distortion limitations.

Thus modifying the channel may be a far more efficient means of ensuring transmission of an important message than encoding that message in a "natural" language which maximizes the rate of transmission of information on a fixed channel.

We have examined the two limits in which either the distributions of $P(Y)$ or of $P(X)$ are kept fixed. The first provides the usual Shannon Coding Theorem, and the second a tuning theorem variant, i.e., a tunable, retina-like, Rate Distortion Manifold, in the sense of Glazebrook and Wallace (2009).

As described above, this result is essentially similar to Shannon's (1959) observation that evaluating the Rate Distortion Function corresponds to finding a channel that is just right for the source and allowed distortion level.

12.3 Metabolic Constraints

Let $Q(\kappa M) \geq 0, Q(0) = 0$ represent a monotonic increasing function of the intensity measure of available metabolic free energy M, and C be the maximum channel capacity available to the cognitive biological processes of interest. One would expect

$$\hat{H} = \frac{\int_0^C H \exp[-H/Q]dH}{\int_0^C \exp[-H/Q]dH} = \frac{Q[\exp(C/Q) - 1] - C}{\exp(C/Q) - 1} \tag{12.6}$$

κ is an inverse energy intensity scaling constant that may be quite small indeed, a consequence of the typically massive entropic translation losses between the metabolic free energy consumed by the physical processes that instantiate information and any actual measure of that information.

Near $M = 0$, expand Q as a Taylor series, with a first term $Q \approx \kappa M$.

This expression tops out quite rapidly with increases in either C or Q, producing energy—and channel capacity—limited results

$$\hat{H} = Q(\kappa M), C/2 \tag{12.7}$$

Then, expanding Q near zero, the two limiting relations imply

$$\begin{aligned} &Q(\kappa M_{X,Y}) < Q(\kappa M_X) + Q(\kappa M_Y) \rightarrow M_{X,Y} < M_X + M_Y\,, \\ &C_{X,Y} < C_X + C_Y \end{aligned} \tag{12.8}$$

The channel capacity constraint can be parsed further for a noisy Gaussian channel. Then (Cover and Thomas 2006)

$$C = 1/2 \log[1 + \mathcal{P}/\sigma^2] \approx 1/2\mathcal{P}/\sigma^2 \tag{12.9}$$

for small $\mathcal{P}/\sigma^2$, where $\mathcal{P}$ is the "power constraint" such that $E(X^2) < \mathcal{P}$ and σ^2 is the noise variance. Assuming information sources X and Y act on the same scale, so that noise variances are the same and quite large, then, taking $\mathcal{P} = Q(\kappa M)$—channel power is determined by available metabolic free energy—and

$$Q(\kappa M_{X,Y}) < Q(\kappa M_X) + Q(\kappa M_Y).$$

Both limiting inequalities are, then, free energy relations leading to a kind of "reaction canalization" in which a set of lower level cognitive modules consumes less metabolic free energy if information crosstalk among them is permitted than under conditions of individual signal isolation.

12.4 Morse Theory

Morse theory examines relations between analytic behavior of a function—the location and character of its critical points—and the underlying topology of the manifold on which the function is defined. We are interested in a number of such functions, for example, a "free energy" constructed from information source uncertainties on a parameter space and "second order" iterations involving parameter manifolds determining critical behavior. These can be reformulated from a Morse theory perspective. Here we follow closely Pettini (2007).

The essential idea of Morse theory is to examine an n-dimensional manifold M as decomposed into level sets of some function $f : M \to \mathbf{R}$ where $\mathbf{R}$ is the set of real numbers. The a-level set of f is defined as

$$f^{-1}(a) = \{x \in M : f(x) = a\},$$

the set of all points in M with $f(x) = a$. If M is compact, then the whole manifold can be decomposed into such slices in a canonical fashion between two limits, defined by the minimum and maximum of f on M. Let the part of M below a be defined as

$$M_a = f^{-1}(-\infty, a] = \{x \in M : f(x) \leq a\}.$$

These sets describe the whole manifold as a varies between the minimum and maximum of f.

Morse Functions are defined as a particular set of smooth functions $f : M \to \mathbf{R}$ as follows. Suppose a function f has a critical point x_c, so that the derivative $df(x_c) = 0$, with critical value $f(x_c)$. Then f is a Morse Function if its critical points are nondegenerate in the sense that the Hessian matrix of second derivatives at x_c, whose elements, in terms of local coordinates, are

$$H_{i,j} = \partial^2 f / \partial x^i \partial x^j,$$

has rank n, which means that it has only nonzero eigenvalues, so that there are no lines or surfaces of critical points and, ultimately, critical points are isolated.

The index of the critical point is the number of negative eigenvalues of H at x_c.

A level set $f^{-1}(a)$ of f is called a critical level if a is a critical value of f, that is, if there is at least one critical point $x_c \in f^{-1}(a)$.

Again following Pettini (2007), the essential results of Morse theory are:

1. If an interval $[a, b]$ contains no critical values of f, then the topology of $f^{-1}[a, v]$ does not change for any $v \in (a, b]$. Importantly, the result is valid even if f is not a Morse Function, but only a smooth function.
2. If the interval $[a, b]$ contains critical values, the topology of $f^{-1}[a, v]$ changes in a manner determined by the properties of the matrix H at the critical points.
3. If $f : M \to \mathbf{R}$ is a Morse Function, the set of all the critical points of f is a discrete subset of M, i.e., critical points are isolated. This is Sard's Theorem.
4. If $f : M \to \mathbf{R}$ is a Morse Function, with M compact, then on a finite interval $[a, b] \subset \mathbf{R}$, there is only a finite number of critical points p of f such that $f(p) \in [a, b]$. The set of critical values of f is a discrete set of $\mathbf{R}$.
5. For any differentiable manifold M, the set of Morse Functions on M is an open dense set in the set of real functions of M of differentiability class r for $0 \leq r \leq \infty$.
6. Some topological invariants of M, that is, quantities that are the same for all the manifolds that have the same topology as M, can be estimated and sometimes computed exactly once all the critical points of f are known: Let the Morse numbers $\mu_i (i = 1, \ldots, m)$ of a function f on M be the number of critical points of f of index i (the number of negative eigenvalues of H). The Euler characteristic of the complicated manifold M can be expressed as the alternating sum of the Morse numbers of any Morse Function on M,

$$\chi = \sum_{i=0}^{m} (-1)^i \mu_i.$$

The Euler characteristic reduces, in the case of a simple polyhedron, to

$$\chi = V - E + F$$

where V, E, and F are the numbers of vertices, edges, and faces in the polyhedron.
7. Another important theorem states that, if the interval $[a, b]$ contains a critical value of f with a single critical point x_c, then the topology of the set M_b defined above differs from that of M_a in a way which is determined by the index, i, of the

critical point. Then M_b is homeomorphic to the manifold obtained from attaching to M_a an i-handle, i.e., the direct product of an i-disk and an $(m-i)$-disk.

Again, Pettini (2007) contains both mathematical details and further references. See, for example, Matsumoto (2002).

12.5 An RDT Proof of the DRT

The Rate Distortion Theorem of information theory asks how much a signal can be compressed and have average distortion, according to an appropriate measure, less than some predetermined limit $D > 0$. The result is an expression for the minimum necessary channel capacity, R, as a function of D. See Cover and Thomas (2006) for details. Different channels have different expressions. For the Gaussian channel under the squared distortion measure,

$$R(D) = \frac{1}{2}\log\left[\frac{\sigma^2}{D}\right] \; D < \sigma^2$$
$$R(D) = 0 \; D \geq \sigma^2 \tag{12.10}$$

where σ^2 is the variance of channel noise having zero mean.

Our concern is how a control signal u_t is expressed in the system response x_{t+1}. We suppose it possible to deterministically retranslate an observed sequence of system outputs $x_1, x_2, x_3, \ldots$ into a sequence of possible control signals $\hat{u}_0, \hat{u}_1, \ldots$ and to compare that sequence with the original control sequence $u_0, u_1, \ldots$, with the difference between them having a particular value under the chosen distortion measure, and hence an observed average distortion.

The correspondence expansion is as follows.

Feynman (2000), following ideas of Bennett, identifies information as a form of free energy. Thus $R(D)$, the minimum channel capacity necessary for average distortion D, is also a free energy measure, and we may define an entropy S as

$$S \equiv R(D) - DdR/dD \tag{12.11}$$

For a Gaussian channel under the squared distortion measure,

$$S = 1/2\log[\sigma^2/D] + 1/2 \tag{12.12}$$

Other channels will have different expressions.

The simplest dynamics of such a system are given by a nonequilibrium Onsager equation in the gradient of S (de Groot and Mazur 1984) so that

$$dD/dt = -\mu dS/dD = \frac{\mu}{2D} \tag{12.13}$$

By inspection,

$$D(t) = \sqrt{\mu t} \tag{12.14}$$

which is the classic outcome of the diffusion equation. For the "natural" channel having $R(D) \propto 1/D$, $D(t) \propto$ the cube root of t.

This correspondence reduction allows an expansion to more complicated systems, in particular, to the control system of Fig. 4.1.

Let $\mathcal{H}$ be the rate at which control information is fed into an inherently unstable control system, in the presence of a further source of control system noise β, in addition to the channel noise defined by σ^2. The simplest generalization of Eq. (12.13), for a Gaussian channel, is the stochastic differential equation

$$dD_t = \left[\frac{\mu}{2D_t} - M(\mathcal{H})\right]dt + \beta D_t dW_t \tag{12.15}$$

where dW_t represents white noise and $M(\mathcal{H}) \geq 0$ is a monotonically increasing function.

This equation has the nonequilibrium steady state expectation

$$D_{nss} = \frac{\mu}{2M(\mathcal{H})} \tag{12.16}$$

measuring the average distortion between what the control system wants and what it gets. In a sense, this is a kind of converse to the famous radar equation which states that a returned signal will be proportional to the inverse fourth power of the distance between the transmitter and the target. But there is a deeper result, leading to the DRT.

Applying the Ito chain rule to Eq. (12.15) (Protter 1990; Khashminskii 2012), it is possible to calculate the expected variance in the distortion as $E(D_t^2) - (E(D_t))^2$. But application of the Ito rule to D_t^2 shows that *no real number solution for its expectation is possible unless the discriminant of the resulting quadratic equation is* ≥ 0, so that a necessary condition for stability is

$$\begin{aligned} M(\mathcal{H}) &\geq \beta\sqrt{\mu} \\ \mathcal{H} &\geq M^{-1}(\beta\sqrt{\mu}) \end{aligned} \tag{12.17}$$

where the second expression follows from the monotonicity of M.

As a consequence of the correspondence reduction leading to Eq. (12.15), we have generalized the DRT of Eq. (4.2). Different "control channels," with different forms of $R(D)$, will give different detailed expressions for the rate of generation of "topological information" by an inherently unstable system.

12.6 An Information Black–Scholes Model

We look at $\mathcal{H}(\rho)$ as the control information rate "cost" of stability at the integrated environmental insult ρ. To determine the mathematical form of $\mathcal{H}(\rho)$ under conditions of volatility, i.e., variability proportional to a signal, we must first model the variability of ρ, most simply taken as

$$d\rho_t = g(t, \rho_t)dt + b\rho_t dW_t \tag{12.18}$$

Here, dW_t is white noise and—counterintuitively—the function $g(t, \rho)$ will fall out of the calculation on the assumption of certain regularities.

$\mathcal{H}(\rho_t, t)$ is the minimum needed incoming rate of control information under the Data Rate Theorem. Expand $\mathcal{H}$ in ρ *using the Ito chain rule* (Protter 1990):

$$\begin{aligned} d\mathcal{H}_t = &\left[\partial\mathcal{H}/\partial t + g(\rho_t, t)\partial\mathcal{H}/\partial\rho + \frac{1}{2}b^2\rho_t^2\partial^2\mathcal{H}/\partial\rho^2\right]dt \\ &+[b\rho_t\partial\mathcal{H}/\partial\rho]dW_t \end{aligned} \tag{12.19}$$

It is now possible to define a Legendre transform, L, of the rate $\mathcal{H}$, by convention having the form

$$L = -\mathcal{H} + \rho\partial\mathcal{H}/\partial\rho \tag{12.20}$$

$\mathcal{H}$ is an information index, a free energy measure in the sense of Feynman (2000), so that L is a classic entropy measure.

We make an approximation, replacing dX with ΔX and applying Eq. (12.19), so that

$$\Delta L = \left(-\partial\mathcal{H}/\partial t - \frac{1}{2}b^2\rho^2\partial^2\mathcal{H}/\partial\rho^2\right)\Delta t \tag{12.21}$$

According to the classical Black–Scholes model (Black and Scholes 1973), the terms in g and dW_t "cancel out," and white noise has been subsumed into the Ito correction factor, a regularity assumption making this an exactly solvable but highly approximate model.

The conventional Black–Scholes calculation takes $\Delta L/\Delta T \propto L$. At nonequilibrium steady state, by some contrast, we can assume $\Delta L/\Delta t = \partial\mathcal{H}/\partial t = 0$, giving

$$-\frac{1}{2}b^2\rho^2\partial^2\mathcal{H}/\partial\rho^2 = 0 \tag{12.22}$$

so that

$$\mathcal{H} = \kappa_1\rho + \kappa_2 \tag{12.23}$$

The κ_i will be nonnegative constants.

12.7 Estimating the Quadratic Variation from Data

The so-called white noise has quadratic variation $\propto t$. The "colored" noise relation can be estimated from the observed periodogram using the methods of Dzhaparidze and Spreij (1994).

For a stochastic process X_t and a finite stopping time T and each real number λ, the *periodogram* of X evaluated at T is defined as

$$I_T(X;\lambda) \equiv |\int_0^T \exp[i\lambda t]dX_t|^2 \tag{12.24}$$

Take ϵ as a real random variable that has a density ω symmetric around zero and consider, for any positive real number L, the quantity

$$E_\epsilon[I_T(X;L\epsilon)] = \int_{-\infty}^{+\infty} I_T(X;Ls)\omega(s)ds \tag{12.25}$$

Dzhaparidze and Spreij (1994) show that, for $L \to \infty$,

$$E_\epsilon[I_T(X;L\epsilon)] \to [X_T, X_T] \tag{12.26}$$

Thus, the quadratic variation can be statistically estimated from observational time series data, as is routinely done in financial engineering, from which, in fact, this analysis is taken.

12.8 A Metabolic Black–Scholes Model

Suppose metabolic free energy to be available at a rate M, and let $R(D)$ be a general RDF for a process ultimately fueled by M. How are M and R related under conditions of volatility? Let

$$dR_t = f(t, R_t)dt + bR_t dW_t \tag{12.27}$$

Let $M(R_t, t)$ be the incoming rate of metabolic free energy, and expand using the Ito chain rule

$$dM_t = \left[\partial M/\partial t + f(R_t, t)\partial M/\partial R + \frac{1}{2}b^2 R_t^2 \partial^2 M/\partial R^2\right]dt$$
$$+[bR_t \partial M/\partial R]dW_t \tag{12.28}$$

We define a quantity L as a Legendre transform of the rate M, by convention having the form

$$L = -M + R\partial M/\partial R \tag{12.29}$$

Again, heuristically, replacing dX with ΔX in these expressions and applying Eq. (12.28) gives

$$\Delta L = \Big(- \partial M/\partial t - \frac{1}{2} b^2 R^2 \partial^2 M/\partial R^2 \Big) \Delta t \tag{12.30}$$

Again, the terms in f and dW_t cancel out, and the effects of noise are subsumed into the Ito correction factor, powerful regularity assumptions that make this an exactly solvable approximate model.

The conventional Black–Scholes calculation takes $\Delta L/\Delta T \propto L$. Here, at nonequilibrium steady state, we assume $\Delta L/\Delta t = C \geq 0$, $\partial M/\partial t = 0$, so that

$$- \frac{1}{2} b^2 R^2 \partial^2 M/\partial R^2 = C \tag{12.31}$$

with solution

$$M = \frac{2C}{b^2} \log[R] + \kappa_1 R + \kappa_2 \tag{12.32}$$

References

Black, F., and M. Scholes. 1973. The pricing of options and corporate liabilities. *Journal of Political Economy* 81: 637–654.

Brown, R. 1987. From groups to groupoids: A brief survey. *Bulletin of the London Mathematical Society* 19: 113–134.

Cannas DaSilva, A., and A. Weinstein. 1999. *Geometric Models for Noncommutative Algebras*. Providence, RI: American Mathematical Society.

Cover, T., and J. Thomas. 2006. *Elements of Information Theory*, 2nd ed. New York: Wiley.

de Groot, S., and P. Mazur. 1984. *Nonequilibrium Thermodynamics*. New York: Dover.

Dzhaparidze, K., and P. Spreij. 1994. Spectral characterization of the optional quadratic variation. *Stochastic Processes and Their Applications* 54: 165–174.

Feynman, R. 2000. *Lectures on Computation*. Boulder: Westview Press.

Glazebrook, J.F., and R. Wallace. 2009. Rate distortion manifolds as model spaces for cognitive information. *Informatica* 33: 309–346.

Golubitsky, M., and I. Stewart. 2006. Nonlinear dynamics of networks: The groupoid formalism. *Bulletin of the American Mathematical Society* 43: 305–364.

Khashminskii, R. 2012. *Stochastic Stability of Differential Equations*, 2nd ed. New York: Springer.

Matsumoto, Y. 2002. *An Introduction to Morse Theory*. Providence, RI: American Mathematical Society.

Pettini, M. 2007. *Geometry and Topology in Hamiltonian Dynamics and Statistical Mechanics*. New York: Springer.

Protter, P. 1990. *Stochastic Integration and Differential Equations*. New York: Springer.

Shannon, C. 1959. Coding theorems for a discrete source with a fidelity criterion. Institute of Radio Engineers International Convention Record, vol. 7, 142–163.

Weinstein, A. 1996. Groupoids: Unifying internal and external symmetry. *Notices of the American Mathematical Association* 43: 744–752.

Index

R. Wallace, *Computational Psychiatry*, DOI 10.1007/978-3-319-53910-2

MIX
Papier aus verantwortungsvollen Quellen
Paper from responsible sources
FSC® C105338

If you have any concerns about our products,
you can contact us on
ProductSafety@springernature.com

In case Publisher is established outside the EU,
the EU authorized representative is:
Springer Nature Customer Service Center GmbH
Europaplatz 3, 69115 Heidelberg, Germany

Printed by Libri Plureos GmbH
in Hamburg, Germany